Nutrition, Diet and Food Allergy

Nutrition, Diet and Food Allergy

Editor

Carla Mastrorilli

MDPI • Basel • Beijing • Wuhan • Barcelona • Belgrade • Manchester • Tokyo • Cluj • Tianjin

Editor
Carla Mastrorilli
Pediatric Hospital Giovanni
XXIII
Italy

Editorial Office
MDPI
St. Alban-Anlage 66
4052 Basel, Switzerland

This is a reprint of articles from the Special Issue published online in the open access journal *Nutrients* (ISSN 2072-6643) (available at: https://www.mdpi.com/journal/nutrients/special_issues/Nutrition_Diet_Food_Allergy).

For citation purposes, cite each article independently as indicated on the article page online and as indicated below:

LastName, A.A.; LastName, B.B.; LastName, C.C. Article Title. *Journal Name* **Year**, *Volume Number*, Page Range.

ISBN 978-3-0365-5479-2 (Hbk)
ISBN 978-3-0365-5480-8 (PDF)

Cover image courtesy of Carla Mastrorilli

Contents

About the Editor

Carla Mastrorilli

Carla Mastrorilli Graduated in Medicine and Surgery, specialist in Pediatrics, PhD in Medical Sciences at University of Parma. Medical Doctor at the Emergency and Pediatrics, Pediatric Hospital Giovanni XXIII, University Hospital of Polyclinic, Bari, Italy. Expert in Allergy and Clinical Immunology, Post-Graduate Diploma in Human Nutrition and Diet. Winner of four scholarships: "Pioneers of Pediatrics" Award of Italian Society of Pediatrics (SIP), "Mentorship Program Award" and "Research Fellowship" of European Academy of Allergy and Clinical Immunology, "Research Award" of the Medical Association of Bari. Scientific projects abroad at Dpt. of Pathophysiology and Allergy Research, Medical University of Vienna and Dpt. of Pediatric Pneumology and Immunology, Charité Medical University, Berlin. Member of Rare allergic diseases and Diagnostic Commissions of Italian Society of Pediatric Allergology and Immunology (SIAIP); Member of Acta BioMedica Editorial Board; Junior Councilor of SIP Puglia section. Peer-reviewer of numerous scientific journals. Titleholder of several scientific publications including: 59 scientific articles, 1 editorial, 2 book chapters, and 2 volumes. Focus of scientific interest: management of allergic diseases in children; administration of digital tools for medical diagnostics; promotion of correct lifestyles and nutrition education.

Preface to "Nutrition, Diet and Food Allergy"

Dietary modules and nutritional habits are fundamental to the growth of a child and ideal performance of the immune system during the first 1000 days of life. A modulation in both innate and adaptive immunity, shaping allergy development, is accomplished by the gut and skin microbiota. Interestingly, bacterial community structures have been found to be different among children affected by food allergies compared to healthy children.

Globally, food allergies affect 1.5% of adults and 5% of children, and this prevalence is increasing in recent decades, representing a public health problem. Different mechanisms are involved in food-allergic diseases with distinctive clinical characteristics: IgE-mediated and non-IgE-mediated phenotypes will be distinguished in the present Special Issue. In the first year of life, cow's milk allergy (CMA) represents the most common allergy. A focal point will be put on peculiar phenotypes of non-IgE mediated CMA, such as food-protein-induced enterocolitis syndrome and Heiner syndrome. Later in childhood, nut and fruit allergies taks hold, and this will be examined in this Special Issue by diagnostic work-up, as well as clinical features.

The management of these fascinating diseases will be discussed with particular attention on nutritional hazards, risks of allergic reactions to new allergens, and problems with mislabelling (precautionary allergen labelling (PAL)). Moreover, the dietary restrictions and the re-introduction of allergens lead to a significant burden for affected patients, fear of accidental ingestions, and related risk of severe reactions, resulting in a reduced quality of life among food-allergic patients. In particular, policies assumed in schools on food allergy will be inquired, focusing on management practices.

I really appreciated the opportunity of taking part of this remarkable work, and I am very thankful to all authors for their support.

Carla Mastrorilli
Editor

nutrients

Article

Maternal Vegetable and Fruit Consumption during Pregnancy and Its Effects on Infant Gut Microbiome

Hsien-Yu Fan [1,2,†], Yu-Tang Tung [3,4,†], Yu-Chen S. H. Yang [5], Justin BoKai Hsu [6], Cheng-Yang Lee [7], Tzu-Hao Chang [7,8], Emily Chia-Yu Su [8,9], Rong-Hong Hsieh [10] and Yang-Ching Chen [1,3,10,11,*]

[1]   Department of Family Medicine, Taipei Medical University Hospital, Taipei 110, Taiwan; d07849004@ntu.edu.tw
[2]   Institute of Epidemiology and Preventive Medicine, National Taiwan University, Taipei 100, Taiwan
[3]   Graduate Institute of Metabolism and Obesity Sciences, College of Nutrition, Taipei Medical University, Taipei 110, Taiwan; peggytung@nchu.edu.tw
[4]   Graduate Institute of Biotechnology, National Chung Hsing University, Taichung 402, Taiwan
[5]   Joint Biobank, Office of Human Research, Taipei Medical University, Taipei 110, Taiwan; can_0131@tmu.edu.tw
[6]   Department of Medical Research, Taipei Medical University Hospital, Taipei 110, Taiwan; justin.bokai@gmail.com
[7]   Office of Information Technology, Taipei Medical University, Taipei 110, Taiwan; nathanlee@tmu.edu.tw (C.-Y.L.); kevinchang@tmu.edu.tw (T.-H.C.)
[8]   Graduate Institute of Biomedical Informatics, College of Medical Science and Technology, Taipei Medical University, Taipei 110, Taiwan; emilysu@tmu.edu.tw
[9]   Clinical Big Data Research Center, Taipei Medical University Hospital, Taipei 110, Taiwan
[10] School of Nutrition and Health Sciences, College of Nutrition, Taipei Medical University, Taipei 110, Taiwan; hsiehrh@tmu.edu.tw
[11] Department of Family Medicine, School of Medicine, College of Medicine, Taipei Medical University, Taipei 110, Taiwan
*   Correspondence: melisa26@tmu.edu.tw or melisa26@gmail.com; Tel.: +886-2-2737-2181 (ext. 3032); Fax: +886-2-2738-9804
†   These authors contributed equally to this study.

Citation: Fan, H.-Y.; Tung, Y.-T.; Yang, Y.-C.S.H.; Hsu, J.B.; Lee, C.-Y.; Chang, T.-H.; Su, E.C.-Y.; Hsieh, R.-H.; Chen, Y.-C. Maternal Vegetable and Fruit Consumption during Pregnancy and Its Effects on Infant Gut Microbiome. *Nutrients* 2021, 13, 1559. https://doi.org/10.3390/nu13051559

Academic Editor: Carla Mastrorilli

Received: 16 March 2021
Accepted: 27 April 2021
Published: 5 May 2021

Publisher's Note: MDPI stays neutral with regard to jurisdictional claims in published maps and institutional affiliations.

Abstract: Maternal nutrition intake during pregnancy may affect the mother-to-child transmission of bacteria, resulting in gut microflora changes in the offspring, with long-term health consequences in later life. Longitudinal human studies are lacking, as only a small amount of studies showing the effect of nutrition intake during pregnancy on the gut microbiome of infants have been performed, and these studies have been mainly conducted on animals. This pilot study explores the effects of high or low fruit and vegetable gestational intake on the infant microbiome. We enrolled pregnant women with a complete 3-day dietary record and received postpartum follow-up. The 16S rRNA gene sequence was used to characterize the infant gut microbiome at 2 months ($n = 39$). Principal coordinate analysis ordination revealed that the infant gut microbiome clustered differently for high and low maternal fruit and vegetable consumption ($p < 0.001$). The linear discriminant analysis effect size and feature selection identified 6 and 17 taxa from both the high and low fruit and vegetable consumption groups. Among the 23 abundant taxa, we observed that six maternal intake nutrients were associated with nine taxa (e.g., *Erysipelatoclostridium*, *Isobaculum*, Lachnospiraceae, Betaproteobacteria, Burkholderiaceae, *Sutterella*, Clostridia, Clostridiales, and *Lachnoclostridium*). The amount of gestational fruit and vegetable consumption is associated with distinct changes in the infant gut microbiome at 2 months of age. Therefore, strategies involving increased fruit and vegetable consumption during pregnancy should be employed for modifying the gut microbiome early in life.

Keywords: nutrients; infant gut microbiome; pregnancy; vegetables; fruits

1. Introduction

According to the Development Origins of Health and Disease (DOHaD) hypothesis, maternal nutrition in pregnancy has a significant impact on offspring disease risk in the future [1]. A maternal diet rich in fruits and vegetables during pregnancy is associated with a reduced risk of allergic diseases [2], and an increased risk of obesity [3]. Given that nutrient intake strongly influences microbiome function and relative abundance, the infant gut microbiome might be a potential mediating factor linking gestational nutritional exposure and future childhood diseases [4].

Maternal nutrition during pregnancy may affect the mother-to-child transmission of bacteria, resulting in gut microflora changes in the child, with long-term consequences after birth [4]. However, evidence supporting the effect of maternal nutrition during pregnancy on the infant gut microbiome remains scarce, and most relevant studies have been conducted on animal models. Chu et al. discovered that a high-fat maternal diet during gestation shapes the offspring gut microbiome in animals (Japanese Macaque) [5] and humans [6]. The mother's high-fat diet was shown to damage the microbiome and immune system of their offspring [7]. The offspring of mothers who consume Western diets displayed a significantly increased effect of Pachycephalospora on Bacteroides, and the microbiome of the offspring of mothers who were fed a high-fat diet had an increased ability to extract energy from the diet. Using a sow model, Li et al. reported that maternal dietary fiber intake alters offspring gut microbiome composition [8]. Similarly, maternal fruit intake was associated with an increased risk of a high *Streptococcus/Clostridium* gut microbiome composition among vaginally delivered infants [9]. Possible mechanisms for the effects of maternal diet during pregnancy on the infant gut microbiome include the transmission of nutrients through amniotic fluid, vaginal delivery, or the placenta. However, the effect of gestational intake of high or low fruit and vegetable intake on the infant microbiome remains unclear in the study of humans.

Several studies suggest that supplementation with nutrients found in fruits and vegetables, such as dietary fiber, vitamin C, and fructose, could modulate the structure of host gut microbes [10]. According to a previous study by Alison et al., a high-fiber diet alters gut microbial ecology and causes significant perturbations at the phylum level [11]. Li et al. found that vitamin C could strongly modulate the gut microbiota [12]. In another animal study, the maternal diet supplemented with fructose appeared to regulate the maternal microbiome significantly, causing infant gut dysbiosis [13]. The intake of maternal dietary fruits and vegetables may not only have an effect on the host, but also on their offspring [14,15]. However, this relationship has, to date, been poorly understood.

In this study, we (1) explored the impact of high/low gestational intakes of fruits and vegetables on the infant microbiome, and (2) investigated the interrelationships between maternal nutrients and the abundance of infant gut microbiome taxa.

2. Materials and Methods

2.1. Cohort Establishment and Data Collection

In July 2018, we formed a Taipei Mother–Infant Nutrition Cohort, which was approved by the Joint Institutional Review Board of Taipei Medical University (N201811050). The 1008 pregnant women in the initial cohort were enrolled from three hospitals and four obstetric clinics, between 2018 and 2019, and they all provided informed consent for this study (Figure S1). Pregnant women with severe diseases (e.g., heart diseases) were excluded. Initially, we included 479 participants who completed the baseline survey and provided dietary data, and 199 of them went on to participate in newborn follow-up visits. We excluded participants who were unwilling to provide infant stool samples in this study. By the end of April 2020, 39 infant stool samples were obtained at 2 months postpartum during home visits.

2.2. Maternal Dietary Assessment

All participants received assistance in installing an image-based dietary assessment application on their smartphones (Cofit Pro version 1.0.0, Taipei, Taiwan). We previously proved the validity and reliability of this image-based dietary assessment application for assessing macronutrients and micronutrients [16]. Registered dietitians taught the participants to use the application on-site for 20–30 min. Participants used their smartphones to take photos of all the food that they consumed for ≥3 days. Dietary records from three matching days (two weekdays and one weekend day) were used for this analysis. Dietary variables were calculated as individual means of the 3-day dietary records. After food record collection, trained dietitians disaggregated the foods into their constituent ingredients, including macronutrients (measured in grams) and micronutrients (measured in milligrams). The macronutrients included carbohydrates and dietary fiber, fats, proteins, and fluids. The micronutrients included vitamins and minerals.

Fruits and vegetables contribute the most to dietary fiber intake in the Taiwanese population; therefore, the 2018 Dietary Guidelines of Taiwan for pregnant women recommend the consumption of 5–9 cups of fruits and vegetables per day. More than 80% of women in our cohort consumed fewer fruits and vegetables than this recommendation during pregnancy. On average, they consumed 4.9 g/day of dietary fiber, well below the recommended 25 g/day. The mean cups of fruits and vegetables were estimated as follows: five cups of fruits and vegetables (the minimum recommended), multiplied by 20% (derived from the average 4.9 g/day divided by the recommended 25 g/day of dietary fiber). The high or low consumption of vegetables and fruit was determined based on more than one cup of fruits and more than one cup of vegetables per day.

2.3. Sample Collection and DNA Extraction

Stool samples were collected during home visits when the infants were 2 months of age. Before home visits, we mailed the participants stool sample collection tubes, which contained commercial deoxyribonucleic acid (DNA) stabilization buffer, to protect DNA from degradation after collection. DNA stabilization contained RNAlater, which protects DNA from degradation at room temperature from days to weeks [17]. The bacterial DNA was extracted using a Qiagen DNA Mini Kit (Qiagen, Hilden, Germany) and stored at $-80\,^{\circ}\mathrm{C}$.

2.4. Targeted 16S rRNA Gene Sequencing

The analytical methods of 16S rDNA analysis were established in a previous study [18]. By referencing Illumina's recommended protocols (https://support.illumina.com/downloads/16s_metagenomic_sequencing_library_preparation.html; accessed on 10 December 2020), we performed library construction and amplification of the 16S rRNA gene. In summary, we used the forward and reverse primers 341F and 805R with Illumina overhang adapter sequencing to amplify the V3–V4 region of the bacterial 16S rRNA gene. A Nextera XT Index kit was then used to adjust the dual-index barcodes to the targets in the amplicon and the Illumina sequencing adapters. The quantity and quality of data in the sequenced library were assessed using a QSep100 analyzer (BiOptic, Taipei, Taiwan). Moreover, by using a MiSeq Reagent kit v3, high-throughput sequencing was performed on an Illumina MiSeq 2000 sequencer.

The bioinformatics analytical process was conducted following the workflow described by Callahan et al. [19]. First, by using the R package DADA2 (v 1.14.1), the filtered reads were managed. Taxonomy assignment was administered using the SILVA database (v128) with a minimum bootstrap confidence level of 80 [20]. Multiple sequence alignment of the structural variants was processed with DECIPHER (v2.14.0), and a phylogenetic tree was built from the alignment using phangorn (v2.5.5) [21]. The count table, taxonomy assignment results, and phylogenetic tree were consolidated into a phyloseq object, and community analyses were created by phyloseq (v1.30.0) [22]. One-way ANOVA followed by the Bonferroni post hoc test were utilized to handle multiple comparison analysis. The analytical process of alpha-diversity and beta-diversity are listed below. The phyloseq package was used to calculate the alpha-diversity. For beta-diversity, principal coordinate analysis (PCoA) was performed on UniFrac distances, and the adonis and betadisper functions from the vegan package (v2.5.6) were used to analyze the dissimilarity of composition among high- and low-consumption groups. The groups were compared with $\alpha = 0.05$ (Kruskal–Wallis and Wilcoxon tests). The UniFrac package (v1.1) was used to compare the community dissimilarity between groups, demonstrated as UniFrac distances [23]. GraPhlAn [24] helped us to perform the enrichment analysis between the groups, which were analyzed using the linear discriminant analysis (LDA), effect size, (LEfSe) method, and a logarithmic LDA score of more than 2 [25] and were then visualized as a cladogram.

2.5. Confounding Factors

We conducted surveys once during pregnancy and twice after birth. The questionnaire administered to the mothers included demographic data for mothers and children, maternal health and disease status, breastfeeding status, perinatal antibiotics use, and children's health and disease status. Data of potential confounders that would influence the infant gut microbiome, such as delivery mode, gestational age, and gestational weight gain, were also collected.

2.6. Data Analysis

Chi-square and t test were used to examine whether demographic characteristics and maternal nutrients differ between the groups. Linear regression was used to examine the relationship between nutrients and the infant gut microbiome at 2 months of age.

3. Results

3.1. Demographic Characteristics

As presented in Table 1, the percentage intake of vegetables ($p < 0.001$) and fruits ($p = 0.08$) differed significantly between the groups, whereas that of dairy, grain, meat, and fat did not differ significantly between groups (all $p > 0.05$). Other potential confounders, such as gestational age (preterm birth or not), excess gestational weight gain, mode of delivery, and breastfeeding or formula feeding, were also not significantly different (all $p > 0.05$).

Table 1. Characteristics of groups with high and low maternal consumption of fruits and vegetables during pregnancy.

Characteristics	High Consumption * ($n = 13$)		Low Consumption ($n = 26$)		Comparison	
	N or Mean	% or (SD)	N or Mean	% or (SD)	Statistics	p
Maternal age at baseline	34.2	(2.6)	33.5	(4.5)	0.64	0.53
<30	0	0.0%	5	19.2%	2.87	0.32
30–35	7	53.8%	11	42.3%		
≥35	6	46.2%	10	38.5%		
Maternal education level						
Senior high school or below	0	0.0%	1	3.8%	0.67	0.82
College	8	61.5%	17	65.4%		
Graduate school and higher	5	38.5%	8	30.8%		
Family income						
<60,000	5	38.4%	10	38.4%	2.14	0.32
60,000–100,000	5	38.4%	5	19.2%		
>100,000	3	23.1%	11	42.3%		
Maternal history of diseases						
Cardiovascular disease	1	7.6%	0	0.0%	0.12	0.72
Gestational diabetes mellitus	0	0.0%	1	3.8%	0.00	1.00
Hyperthyroidism	2	15.3%	0	0.0%	1.64	0.20
Hypothyroidism	1	7.6%	0	0.0%	0.12	0.72

Table 1. *Cont.*

Characteristics	High Consumption * (*n* = 13)		Low Consumption (*n* = 26)		Comparison	
	N or Mean	% or (SD)	N or Mean	% or (SD)	Statistics	*p*
Timing of dietary assessment						
<13 weeks	0	0.0%	8	30.8%	5.12	0.08
13–28 weeks	8	61.5%	12	46.2%		
>28 weeks	5	38.5%	6	23.1%		
Gestational age	38.7	(1.3)	38.1	(1.2)	1.20	0.24
<37 weeks	1	8.3%	2	8.3%	0.00	1.00
Excess gestational weight gain	2	15.4%	4	15.4%	0.00	1.00
Normal spontaneous delivery	9	69.2%	17	65.4%	0.00	1.00
Antepartum antibiotics	1	7.7%	4	15.4%	0.03	0.64
Group B streptococcus positive	2	15.4%	5	19.2%	0.00	1.00
Neonatal sex (male)	7	58.3%	17	65.4%	0.04	0.72
Breastfeeding (yes)	5	38.5%	10	38.5%	0.00	1.00
Dietary intake						
Calories (kcal)	1679.3	(354.4)	1541.1	(378.9)	204.0	0.31
Vegetables (cups [†])	2.1	(0.7)	1.8	(1.5)	228.0	0.08
Fruit (cups [†])	1.9	(0.5)	0.6	(0.6)	317.0	<0.001
Dairy (cups [†])	0.7	(0.5)	0.6	(0.8)	202.5	0.32
Grain (1/4 cups [†])	8.4	(2.1)	8.2	(2.4)	175.0	0.87
Meat (1/6 cups [†])	5.9	(2.4)	6.0	(3.3)	185.0	0.65
Fat (tbsp [†])	5.5	(1.8)	5.9	(2.3)	150.0	0.59

* High consumption was defined as ≥1 cup of fruits or vegetables per day. [†] One cup = 240 mL; tbsp = tablespoon (5 mL).

3.2. Variation of Maternal Nutrient Intake

Maternal nutrient intake (macronutrients and micronutrients) in the high and low vegetable and fruit consumption groups during pregnancy is presented in Figure 1. The mothers with high fruit and vegetable consumption had a significantly higher intake of macronutrients (glucose, fructose, and dietary fiber), vitamins (folic acid and ascorbic acid), and minerals (potassium) than mothers with low fruit and vegetable consumption.

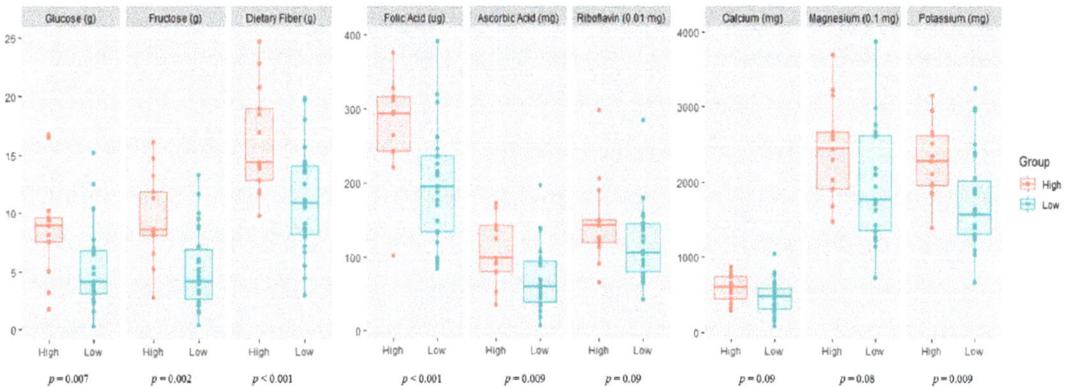

Figure 1. Maternal nutrient intake during pregnancy according to vegetable and fruit consumption. High, high maternal consumption of fruits and vegetables during pregnancy; Low, low maternal consumption of fruits and vegetables during pregnancy.

3.3. Variation of Infant Gut Microbiome

High or low maternal consumption of vegetables or fruits during pregnancy did not affect the alpha diversity of the infant's gut microbiome (Figure S2). To establish the effect of maternal fruit and vegetable intake during pregnancy on the infant's gut microbiome composition, we conducted Illumina-generated 16S rRNA amplicon sequencing from 39 samples. The PCoA based on unweighted UniFrac distances revealed that the microbiome of 2-month-old infants varied depending on whether the maternal consumption of fruits and vegetables gestation was high or low (Figure 2A). However, other potential confounders, such as maternal age, maternal education level, family income, gestational age, excess gestational weight gain, delivery mode, antepartum antibiotics, group B *Streptococcus* positivity, sex of the infant, and breastfeeding did not affect the infant gut microbiome (Figure S3). As shown in Figure 2B, LEfSe revealed that the counts of Propionibacteriales, Propionibacteriaceae, *Cutibacterium*, Tannerellaceae, *Parabacteroides*, and *Lactococcus* were higher in the microbiome of 2-month-old infants with high maternal vegetable and fruit consumption. However, the counts of *Prevotella_2*, *Prevotella_9*, *Isobaculum*, Clostridia, Clostridiales, Lachnospiraceae, *Hungatella*, *Lachnoclostridium*, Ruminococcaceae, *Flavonifractor*, *Erysipelatoclostridium*, Acidaminococcaceae, *Phascolarctobacterium*, *Megamonas*, Betaproteobacteriales, Burkholderiaceae, and Sutterella were higher in the microbiome of 2-months-old infants with low maternal fruit and vegetable consumption.

3.4. Maternal Nutrient Intake and Infant Gut Microbiome

Heatmaps of the correlation between maternal nutrient intake during pregnancy and infant gut microbiome at 2 months of age are displayed in Figure 3. A high-fructose maternal diet was negatively associated with *Erysipelatoclostridium*. A high-glucose maternal diet was significantly associated with an enrichment of *Isobaculum* in the infant gut microbiome. The Lachnospiraceae count was lower among infants with higher maternal consumption of dietary fiber (Figure 3A). Betaproteobacteria, Burkholderiaceae, and *Sutterella* were strongly negatively correlated with folic acid, and Betaproteobacteria and Burkholderiaceae were negatively correlated with ascorbic acid (Figure 3B). Clostridia, Clostridiales, and Lachnospiraceae were negatively correlated with both magnesium and potassium, and *Lachnoclostridium* was negatively correlated with potassium.

(A)

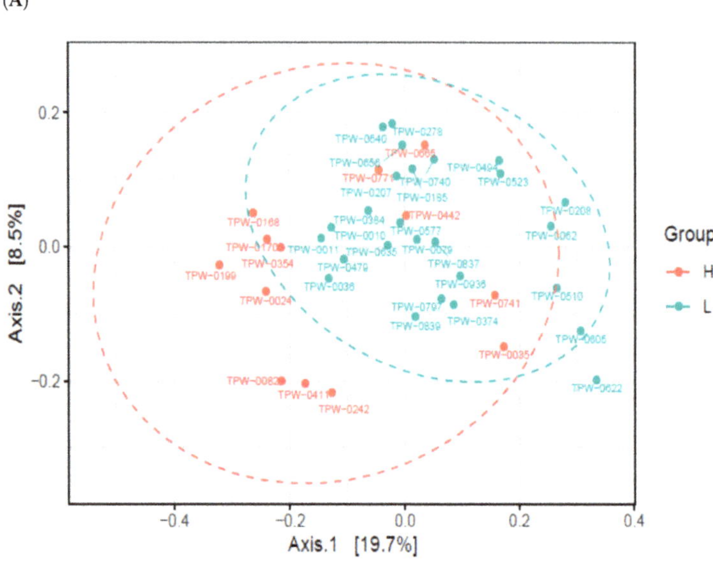

Adonis *p* = 0.001; Betadisper; *p* = 0.90.

Figure 2. *Cont.*

(B)

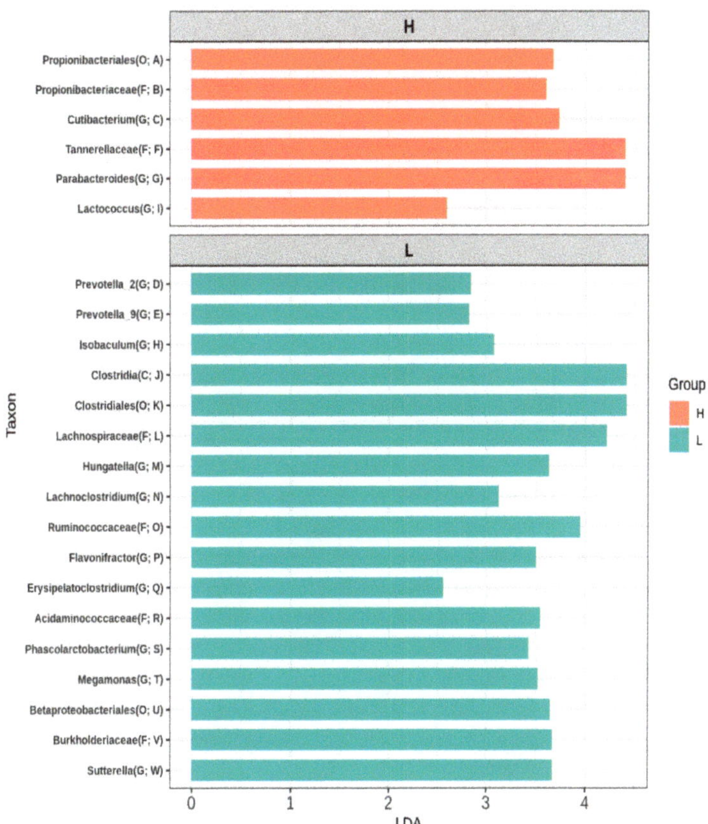

Figure 2. Variations in the infant gut microbiome at 2 months of age according to maternal fruit and vegetable consumption during pregnancy. (**A**) Principal coordinates analysis (PCoA) on unweighted unique fraction (UniFrac). (**B**) linear discriminant analysis effect size (LEfSe). The ordination is from A to W in a tree diagram (Figure S3). Definition of abbreviation: H, high maternal consumption of fruits and vegetables during pregnancy; L, low maternal consumption of fruits and vegetables during pregnancy; O, order; F, family; G, genus.

As shown in Figure S4, counts of *Hungatella* and *Megamonas* were lower among infants with higher maternal consumption of vegetables. Moreover, the count of *Erysipelatoclostridium* was lower among those with higher maternal consumption of fruits, and the count of *Megamonas* was higher among those with higher maternal consumption of dairy. The count of *Isobaculum* was lower among those with higher maternal consumption of grains, whereas the count of *Flavonifractor* was higher. The counts of both Lachnospiraceae and *Lachnoclostridium* were lower among those with higher maternal consumption of meat.

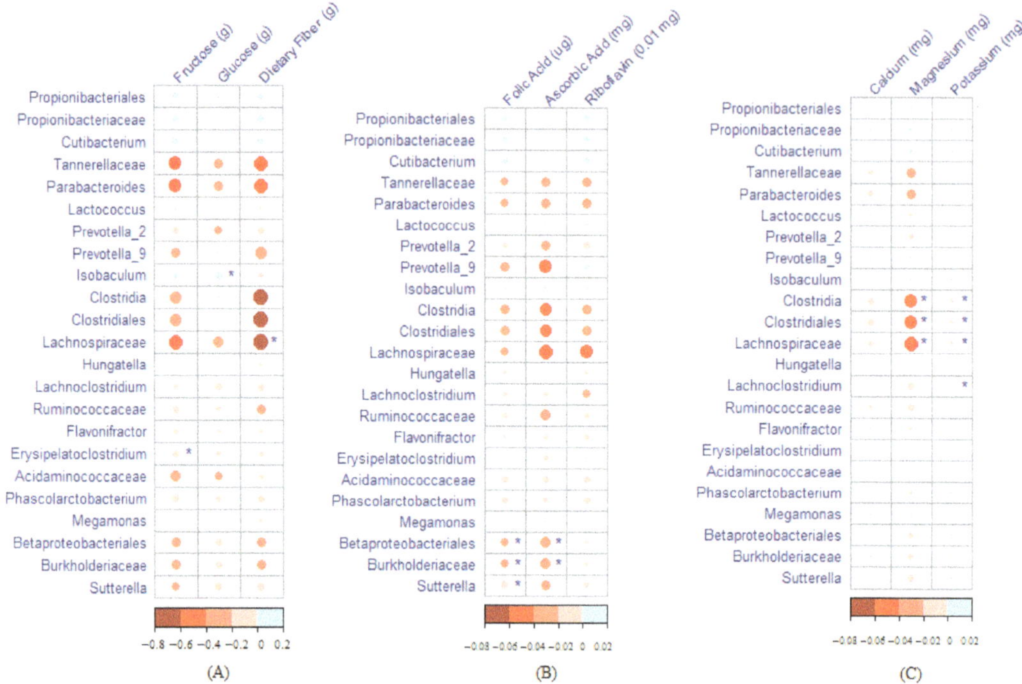

Figure 3. Heatmaps showing the correlation between maternal nutrient intake during pregnancy and the infant gut microbiome at 2 months of age (* Significant association). (**A**) Macronutrients. (**B**) Vitamins. (**C**) Minerals.

4. Discussion

We demonstrated that, in this mother–infant nutrition cohort, the infant gut microbiome at 2 months of age varied according to the level of maternal fruit and vegetable consumption during pregnancy. We identified 6 and 17 taxa in the infant gut microbiome from the high and low fruit and vegetable consumption groups, respectively. Furthermore, we have shown the detailed nutrients and gut microbiome taxonomic interactions.

4.1. Maternal Fruit and Vegetable Consumption Affects Infant Microbiome

High or low maternal fruit and vegetable consumption was significantly correlated with infant gut microbiome composition. Gut microbiome composition is related to the intake of dietary fiber [26], which is fermented by certain bacteria, producing short-chain fatty acids (SCFAs) such as acetate, propionate, and butyrate [27]. Animal studies have reported an association of increased maternal dietary microbiome-accessible fiber and SCFA exposure during pregnancy, with a reduced incidence of asthma in offspring [11,28], and this effect persists into adulthood [11]. A follow-up small human component (*n* = 61) of the same study indicated that an association exists between reduced dietary fiber intake and reduced serum acetate levels in pregnant women. A separate component (*n* = 40) revealed a correlation between serum acetate levels that were lower than the median, and increased frequency of coughing/wheezing during the child's first year of life [11,28].

In a mouse model study, the plasma SCFA levels of the offspring of mice fed a high-fiber diet were higher than those of mice fed a no-fiber diet, and the frequencies of thymic regulatory T cells (Tregs) and peripheral Tregs were higher in the offspring of high-fiber-diet-fed mice [26]. During pregnancy, SCFA (such as acetate) can cross the placenta and affect the expression of fetal lung genes, such as NPPA, which encodes ANP (a molecule related to epithelial biology and immune regulation) [11]. In demonstrating associations between maternal high dietary fiber intake, antenatal exposure to SCFAs, and offspring allergic diseases, these mouse experiments have suggested a possible target for interventions to reduce the burden of allergic diseases; however, no clinical trials have investigated the protective effects of maternal microbiome against allergic diseases through a maternal high-fiber diet.

In the present study, we observed a higher relative abundance of *Cutibacterium*, *Parabacteroides*, and *Lactococcus* in the fecal microbiome of infants exposed to high vegetable and fruit consumption during gestation. However, the higher abundance of *Prevotella_2*, *Prevotella_9*, *Isobaculum*, *Hungatella*, *Lachnoclostridium*, *Flavonifractor*, *Erysipelatoclostridium*, *Phascolarctobacterium*, *Megamonas*, and *Sutterella* was associated with low fruit and vegetable consumption. Our findings regarding the beneficial effects of *Cutibacterium*, *Parabacteroides*, and *Lactococcus* on infant immunity are consistent with those of previous studies. *Cutibacterium* ferments hexoses through the Embden–Meyerhof pathway to produce pyruvate, which is further metabolized into propionate [29], whose consumption was reported to reduce antigen presentation on dendritic cells as a result of GPR41-dependent modulation of hematopoiesis and affected allergic diseases in a mouse model [28]. *Parabacteroides* have multiple beneficial effects on human health. P. distasonis can improve human bowel health [30] and is negatively associated with celiac disease [31]. It can reduce weight gain, hyperglycemia, and liver steatosis in ob/ob and high-fat diet mice [32] and significantly reduce the severity of intestinal inflammation in murine models of acute and chronic colitis [33]. The SCFA-producing *Parabacteroides* were richer in the cecum and colorectum, where, accordingly, more SCFAs were produced [34]. *Lactococcus* lactis activates innate immunity and protects from infections [35,36]. Moreover, some Lactobacilli can produce SCFAs [34], which can induce Tregs to modulate gut immune responses [37–39], and can shape the pulmonary immune environment and influence the severity of allergic inflammation [28].

4.2. Maternal Difference in Six Nutrients and 23 Bacterial Taxa

A comparative study on fecal samples from volunteers with diets low in fructose—a determinant of microbial diversity—revealed that the relative abundance of *Erysipelatoclostridium* was lower among those with a high-fructose syrup diet than among those on a fruit-based diet [40]. The effects of maternal nutrients on the infant gut microbiome have never before been examined in a human model. Here, we highlight that maternal exposure to fructose reduces the abundance of *Erysipelatoclostridium* in the infant gut microbiome. Previous studies on rats have demonstrated that fructose adversely affects intestinal permeability and disrupts the maternal microbiome, leading to altered offspring gut development [13,41]. Fructose may inhibit the growth of harmful flora and promote the growth of beneficial and neutral flora.

Micronutrients may be associated with the abundance of certain taxa in the infant gut microbiome. For example, higher consumption of folate is associated with a lower abundance of Lachnospiraceae [42]. Folate explains 8% of the relative abundance of Lachnospiraceae [42]. Here, we have demonstrated that folate was significantly inversely associated with the abundance of Betaproteobacteria. In the present study, the abundance of Lachnospiraceae was inversely affected by dietary fiber, magnesium, and potassium. However, the effects of micronutrients in the above association in the mother or child remain unclear. Regarding maternal vitamin intake and gut microbiome, the intake of ascorbic acid (vitamin C) during pregnancy was positively correlated with the abundance of *Staphylococcus* [43]. Although the role of ascorbic acid in *Staphylococcus* metabolism remains unclear, both have been linked to the immune profile [44].

4.3. Strengths and Limitations

This is the first human study to demonstrate that low gestational consumption of fruits and vegetables affects the infant gut microbiome. Our data support previous findings from animal studies [11]. Moreover, instead of using a food frequency questionnaire, we used 3-day dietary records to obtain details regarding nutritional intake, thus enabling us to investigate the correlations between gestational nutrition and the abundance of infant gut microbiome taxa.

Our study has some limitations. The generalizability of our findings may be limited by our relatively small sample size. However, the identification of three unhealthy infant gut microbial taxa in the low vegetable/fruit consumption group agrees with the data obtained from other studies conducted in Asian countries [45,46]. Moreover, we were unable to collect samples for evaluating the infant gut microbiome and SCFA at multiple time points. Larger and longer studies that better account for antenatal and postnatal nutritional exposure factors are warranted, to elucidate the detailed mechanisms linking gestational nutritional exposure to early allergic diseases or other chronic diseases.

5. Conclusions

A maternal diet rich in fruits and vegetables during pregnancy may alter the infant gut microbiome. Higher maternal nutritional intake of fructose, dietary fiber, folic acid, and ascorbic acid was negatively associated with the abundance of unhealthy infant gut microbiomes, such as *Erysipelatoclostridium*, Betaproteobacteria, and Lachnospiraceae. Therefore, strategies should be applied for modifying the gut microbiome early in life through the promotion of a higher intake of fruits and vegetables during pregnancy.

Supplementary Materials: The following are available online at https://www.mdpi.com/article/10.3390/nu13051559/s1. Figure S1. Flowchart of participant recruitment; Figure S2. Alpha diversity of the infant gut microbiome according to maternal vegetable and fruit consumption during pregnancy; Figure S3. Variations in the infant gut microbiome are not explained by other potential confounders; Figure S4. Relationship of intake of food from specific food groups to the abundance of specific gut microbiome.

Author Contributions: H.-Y.F. contributed to hypothesis generation, data interpretation, and manuscript writing; Y.-T.T. contributed to data collection, data interpretation and manuscript writing; Y.-C.C. reviewed the study design and data collection and interpretation, supervised the study, and critically revised the manuscript for essential intellectual content. Y.-C.S.H.Y.; J.B.H.; C.-Y.L.; T.-H.C.; E.C.-Y.S.; R.-H.H. assisted in the data collection, statistical analysis, and manuscript review. All authors have read and agreed to the published version of the manuscript.

Funding: This work was supported by the Taipei Medical University Hospital (108TMU-TMUH-01; 109TMU-TMUH-01) and the Ministry of Science and Technology of Taiwan (109-2314-B-038-057, 109-2314-B-038-058).

Institutional Review Board Statement: The Taipei Mother–Infant Nutrition Cohort Study was approved by the Joint Institutional Review Board of Taipei Medical University (N201811050).

Informed Consent Statement: Informed consent was obtained from all subjects involved in the study.

Acknowledgments: We thank the Translational Laboratory, and the Department of Medical Research from Taipei Medical University Hospital for providing support in the preparation of gut microbiota samples. The authors would like to acknowledge the technologic and analysis support provided by TMU (Taipei Medical University) Core Laboratory of Human Microbiome.

Conflicts of Interest: The authors have no competing interests to declare. The funders had no role in the design of the study; in the collection, analyses, or interpretation of data; in the writing of the manuscript; or in the decision to publish the results.

References

1. Barker, D.J. The fetal and infant origins of adult disease. *BMJ* **1990**, *301*, 1111. [CrossRef] [PubMed]
2. Netting, M.J.; Middleton, P.F.; Makrides, M. Does maternal diet during pregnancy and lactation affect outcomes in offspring? A systematic review of food-based approaches. *Nutrients* **2014**, *30*, 1225–1241. [CrossRef]
3. Martin, C.L.; Siega-Riz, A.M.; Sotres-Alvarez, D.; Robinson, W.R.; Daniels, J.L.; Perrin, E.M.; Stuebe, A.M. Maternal Dietary Patterns during Pregnancy Are Associated with Child Growth in the First 3 Years of Life. *J. Nutr.* **2016**, *146*, 2281–2288. [CrossRef]
4. Chu, D.M.; Meyer, K.M.; Prince, A.L.; Aagaard, K.M. Impact of maternal nutrition in pregnancy and lactation on offspring gut microbial composition and function. *Gut Microbes* **2016**, *7*, 459–470. [CrossRef]

5. Ma, J.; Prince, A.L.; Bader, D.; Hu, M.; Ganu, R.; Baquero, K.; Blundell, P.; Harris, R.A.; Frias, A.E.; Grove, K.L.; et al. High-fat maternal diet during pregnancy persistently alters the offspring microbiome in a primate model. *Nat. Commun.* **2014**, *5*, 1–11. [CrossRef] [PubMed]

6. Chu, D.M.; Antony, K.M.; Ma, J.; Prince, A.L.; Showalter, L.; Moller, M.; Aagaard, K.M. The early infant gut microbiome varies in association with a maternal high-fat diet. *Genome Med.* **2016**, *8*, 1–12. [CrossRef] [PubMed]

7. Myles, I.A.; Fontecilla, N.M.; Janelsins, B.M.; Vithayathil, P.J.; Segre, J.A.; Datta, S.K. Parental Dietary Fat Intake Alters Offspring Microbiome and Immunity. *J. Immunol.* **2013**, *191*, 3200–3209. [CrossRef] [PubMed]

8. Li, Y.; Liu, H.; Zhang, L.; Yang, Y.; Lin, Y.; Zhuo, Y.; Fang, Z.; Che, L.; Feng, B.; Xu, S.; et al. Maternal Dietary Fiber Composition during Gestation Induces Changes in Offspring Antioxidative Capacity, Inflammatory Response, and Gut Microbiota in a Sow Model. *Int. J. Mol. Sci.* **2019**, *21*, 31. [CrossRef] [PubMed]

9. Lundgren, S.N.; Madan, J.C.; Emond, J.A.; Morrison, H.G.; Christensen, B.C.; Karagas, M.R.; Hoen, A.G. Maternal diet during pregnancy is related with the infant stool microbiome in a delivery mode-dependent manner. *Microbiome* **2018**, *6*, 1–11. [CrossRef]

10. Yang, Q.; Liang, Q.; Balakrishnan, B.; Belobrajdic, D.P.; Feng, Q.-J.; Zhang, W. Role of Dietary Nutrients in the Modulation of Gut Microbiota: A Narrative Review. *Nutrients* **2020**, *12*, 381. [CrossRef]

11. Thorburn, A.N.; McKenzie, C.I.; Shen, S.; Stanley, D.; Macia, L.; Mason, L.J.; Roberts, L.K.; Wong, C.H.Y.; Shim, R.; Robert, R.; et al. Evidence that asthma is a developmental origin disease influenced by maternal diet and bacterial metabolites. *Nat. Commun.* **2015**, *6*, 7320. [CrossRef]

12. Li, L.; Krause, L.; Somerset, S. Associations between micronutrient intakes and gut microbiota in a group of adults with cystic fibrosis. *Clin. Nutr.* **2017**, *36*, 1097–1104. [CrossRef]

13. Astbury, S.; Song, A.; Zhou, M.; Nielsen, B.; Hoedl, A.; Willing, B.P.; Symonds, M.E.; Bell, R.C. High Fructose Intake During Pregnancy in Rats Influences the Maternal Microbiome and Gut Development in the Offspring. *Front. Genet.* **2018**, *9*, 203. [CrossRef] [PubMed]

14. Dunn, A.B.; Jordan, S.; Baker, B.J.; Carlson, N.S. The Maternal Infant Microbiome: Considerations for Labor and Birth. *MCN Am. J. Matern. Child Nurs.* **2017**, *42*, 318–325. [CrossRef]

15. Wang, S.; Ryan, C.A.; Boyaval, P.; Dempsey, E.M.; Ross, R.P.; Stanton, C. Maternal Vertical Transmission Affecting Early-life Microbiota Development. *Trends Microbiol.* **2020**, *28*, 28–45. [CrossRef]

16. Wang, J.-S.; Hsieh, R.-H.; Tung, Y.-T.; Chen, Y.-H.; Yang, C.; Chen, Y.C. Evaluation of a Technological Image-Based Dietary Assessment Tool for Children during Pubertal Growth: A Pilot Study. *Nutrients* **2019**, *11*, 2527. [CrossRef]

17. Carruthers, L.V.; Moses, A.; Adriko, M.; Faust, C.L.; Tukahebwa, E.M.; Hall, L.J.; Ranford-Cartwright, L.C.; Lamberton, P.H. The impact of storage conditions on human stool 16S rRNA microbiome composition and diversity. *PeerJ* **2019**, *7*, e8133. [CrossRef]

18. Yang, Y.-C.S.H.; Chang, H.-W.; Lin, I.H.; Chien, L.-N.; Wu, M.-J.; Liu, Y.-R.; Chu, P.G.; Xie, G.; Dong, F.; Jia, W.; et al. Long-term Proton Pump Inhibitor Administration Caused Physiological and Microbiota Changes in Rats. *Sci. Rep.* **2020**, *10*, 866. [CrossRef]

19. Callahan, B.J.; Sankaran, K.; Fukuyama, J.A.; McMurdie, P.J.; Holmes, S.P. Bioconductor Workflow for Microbiome Data Analysis: From raw reads to community analyses. *F1000Research* **2016**, *5*, 1492. [CrossRef]

20. Quast, C.; Pruesse, E.; Yilmaz, P.; Gerken, J.; Schweer, T.; Yarza, P.; Peplies, J.; Glöckner, F.O. The SILVA ribosomal RNA gene database project: Improved data processing and web-based tools. *Nucleic Acids Res.* **2013**, *41*, D590–D596. [CrossRef]

21. Schliep, K.P. phangorn: Phylogenetic analysis in R. *Bioinformatics* **2011**, *27*, 592–593. [CrossRef]

22. McMurdie, P.J.; Holmes, S. Phyloseq: An R Package for Reproducible Interactive Analysis and Graphics of Microbiome Census Data. *PLoS ONE* **2013**, *8*, e61217. [CrossRef] [PubMed]

23. Chen, J.; Bittinger, K.; Charlson, E.S.; Hoffmann, C.; Lewis, J.; Wu, G.D.; Collman, R.G.; Bushman, F.D.; Li, H. Associating microbiome composition with environmental covariates using generalized UniFrac distances. *Bioinformatics* **2012**, *28*, 2106–2113. [CrossRef] [PubMed]

24. Asnicar, F.; Weingart, G.; Tickle, T.L.; Huttenhower, C.; Segata, N. Compact graphical representation of phylogenetic data and metadata with GraPhlAn. *PeerJ* **2015**, *3*, e1029. [CrossRef]

25. Segata, N.; Izard, J.; Waldron, L.; Gevers, D.; Miropolsky, L.; Garrett, W.S.; Huttenhower, C. Metagenomic biomarker discovery and explanation. *Genome Biol.* **2011**, *12*, R60. [CrossRef]

26. Nakajima, A.; Kaga, N.; Nakanishi, Y.; Ohno, H.; Miyamoto, J.; Kimura, I.; Hori, S.; Sasaki, T.; Hiramatsu, K.; Okumura, K.; et al. Maternal High Fiber Diet during Pregnancy and Lactation Influences Regulatory T Cell Differentiation in Offspring in Mice. *J. Immunol.* **2017**, *199*, 3516–3524. [CrossRef]

27. Maslowski, K.M.; Mackay, C.R. Diet, gut microbiota and immune responses. *Nat. Immunol.* **2010**, *12*, 5–9. [CrossRef]

28. Trompette, A.; Gollwitzer, E.S.; Yadava, K.; Sichelstiel, A.K.; Sprenger, N.; Ngom-Bru, C.; Blanchard, C.; Junt, T.; Nicod, L.P.; Harris, N.L.; et al. Gut microbiota metabolism of dietary fiber influences allergic airway disease and hematopoiesis. *Nat. Med.* **2014**, *20*, 159–166. [CrossRef]

29. Wood, H.G. Metabolic Cycles in the Fermentation by Propionic Acid Bacteria. *Curr. Top. Cell. Regul.* **1981**, *18*, 255–287. [CrossRef]

30. Clarke, J.M.; Topping, D.L.; Christophersen, C.T.; Bird, A.R.; Lange, K.; Saunders, I.; Cobiac, L. Butyrate esterified to starch is released in the human gastrointestinal tract. *Am. J. Clin. Nutr.* **2011**, *94*, 1276–1283. [CrossRef]

31. Sánchez, E.; Donat, E.; Ribes-Koninckx, C.; Calabuig, M.; Sanz, Y. Intestinal Bacteroides species associated with coeliac disease. *J. Clin. Pathol.* **2010**, *63*, 1105–1111. [CrossRef] [PubMed]

32. Wang, K.; Liao, M.; Zhou, N.; Bao, L.; Ma, K.; Zheng, Z.; Wang, Y.; Liu, C.; Wang, W.; Wang, J.; et al. Parabacteroides distasonis Alleviates Obesity and Metabolic Dysfunctions via Production of Succinate and Secondary Bile Acids. *Cell Rep.* **2019**, *26*, 222–235. [CrossRef] [PubMed]
33. Kverka, M.; Zakostelska, Z.; Klimesova, K.; Sokol, D.; Hudcovic, T.; Hrncir, T.; Rossmann, P.; Mrazek, J.; Kopecny, J.; Verdu, E.F.; et al. Oral administration of Parabacteroides distasonis antigens attenuates experimental murine colitis through modulation of immunity and microbiota composition. *Clin. Exp. Immunol.* **2010**, *163*, 250–259. [CrossRef] [PubMed]
34. Du, X.; Xiang, Y.; Lou, F.; Tu, P.; Zhang, X.; Hu, X.; Lyu, W.; Xiao, Y. Microbial Community and Short-Chain Fatty Acid Mapping in the Intestinal Tract of Quail. *Animals* **2020**, *10*, 1006. [CrossRef]
35. Beck, B.R.; Kim, D.; Jeon, J.; Lee, S.-M.; Kim, H.K.; Kim, O.-J.; Lee, J.I.; Suh, B.S.; Do, H.K.; Lee, K.H.; et al. The effects of combined dietary probiotics Lactococcus lactis BFE920 and Lactobacillus plantarum FGL0001 on innate immunity and disease resistance in olive flounder (Paralichthys olivaceus). *Fish Shellfish Immunol.* **2015**, *42*, 177–183. [CrossRef]
36. Kim, D.; Beck, B.R.; Heo, S.-B.; Kim, J.; Kim, H.D.; Lee, S.-M.; Kim, Y.; Oh, S.Y.; Lee, K.; Do, H.; et al. Lactococcus lactis BFE920 activates the innate immune system of olive flounder (Paralichthys olivaceus), resulting in protection against Streptococcus iniae infection and enhancing feed efficiency and weight gain in large-scale field studies. *Fish Shellfish Immunol.* **2013**, *35*, 1585–1590. [CrossRef]
37. Furusawa, Y.; Obata, Y.; Fukuda, S.; Endo, T.A.; Nakato, G.; Takahashi, D.; Nakanishi, Y.; Uetake, C.; Kato, K.; Kato, T.; et al. Commensal microbe-derived butyrate induces the differentiation of colonic regulatory T cells. *Nature* **2013**, *504*, 446–450. [CrossRef]
38. Arpaia, N.; Campbell, C.; Fan, X.; Dikiy, S.; van der Veeken, J.; deRoos, P.; Liu, H.; Cross, J.R.; Pfeffer, K.; Coffer, P.J.; et al. Metabolites produced by commensal bacteria promote peripheral regulatory T-cell generation. *Nature* **2013**, *504*, 451–455. [CrossRef]
39. Smith, P.M.; Howitt, M.R.; Panikov, N.; Michaud, M.; Gallini, C.A.; Bohlooly-Y, M.; Glickman, J.N.; Garrett, W.S. The Microbial Metabolites, Short-Chain Fatty Acids, Regulate Colonic Treg Cell Homeostasis. *Science* **2013**, *341*, 569–573. [CrossRef]
40. Gonzalez-Granda, A.; Beisner, J.; Basrai, M.; Damms-Machado, A.; Bischoff, S. Fructose-induced intestinal microbiota shift following two types of short-term high fructose dietary phases: A microbiota analysis of a crossover intervention study in healthy women. *Nutrition* **2019**, *44*, 24. [CrossRef]
41. Do, M.H.; Lee, E.; Oh, M.-J.; Kim, Y.; Park, H.-Y. High-Glucose or -Fructose Diet Cause Changes of the Gut Microbiota and Metabolic Disorders in Mice without Body Weight Change. *Nutrition* **2018**, *10*, 761. [CrossRef]
42. Gurwara, S.; Ajami, N.J.; Jang, A.; Hessel, F.C.; Chen, L.; Plew, S.; Wang, Z.; Graham, D.Y.; Hair, C.; White, D.L.; et al. Dietary Nutrients Involved in One-Carbon Metabolism and Colonic Mucosa-Associated Gut Microbiome in Individuals with an Endoscopically Normal Colon. *Nutrition* **2019**, *11*, 613. [CrossRef]
43. Padilha, M.; Danneskiold-Samsøe, N.B.; Brejnrod, A.; Hoffmann, C.; Cabral, V.P.; Iaucci, J.D.M.; Sales, C.H.; Fisberg, R.M.; Cortez, R.V.; Brix, S.; et al. The Human Milk Microbiota is Modulated by Maternal Diet. *Microorganisms* **2019**, *7*, 502. [CrossRef]
44. Hoppu, U.; Rinne, M.; Salo-Väänänen, P.; Lampi, A.-M.; Piironen, V.; Isolauri, E.; Salo-Väävnänen, P. Vitamin C in breast milk may reduce the risk of atopy in the infant. *Eur. J. Clin. Nutr.* **2004**, *59*, 123–128. [CrossRef]
45. Shen, X.; Wang, M.; Zhang, X.; He, M.; Li, M.; Cheng, G.; Wan, C.; He, F. Dynamic construction of gut microbiota may influence allergic diseases of infants in Southwest China. *BMC Microbiol.* **2019**, *19*, 1–13. [CrossRef]
46. Chua, H.-H.; Chou, H.-C.; Tung, Y.-L.; Chiang, B.-L.; Liao, C.-C.; Liu, H.-H.; Ni, Y.-H. Intestinal Dysbiosis Featuring Abundance of Ruminococcus gnavus Associates with Allergic Diseases in Infants. *Gastroenterology* **2018**, *154*, 154–167. [CrossRef]

Article

A Combined Analysis of Gut and Skin Microbiota in Infants with Food Allergy and Atopic Dermatitis: A Pilot Study

Ewa Łoś-Rycharska [1,*], Marcin Gołębiewski [2,3,*], Marcin Sikora [3], Tomasz Grzybowski [4], Marta Gorzkiewicz [4], Maria Popielarz [1], Julia Gawryjołek [1] and Aneta Krogulska [1]

1. Department of Pediatrics, Allergology and Gastroenterology, Collegium Medicum in Bydgoszcz, Nicolaus Copernicus University in Torun, 87-100 Toruń, Poland; maria.popielarz@cm.umk.pl (M.P.); julia.gawryjolek@cm.umk.pl (J.G.); aneta.krogulska@cm.umk.pl (A.K.)
2. Department of Plant Physiology and Biotechnology, Nicolaus Copernicus University in Torun, 87-100 Toruń, Poland
3. Interdisciplinary Centre of Modern Technologies, Nicolaus Copernicus University in Torun, 87-100 Toruń, Poland; sikoramar@op.pl
4. Department of Forensic Medicine, Collegium Medicum in Bydgoszcz, Nicolaus Copernicus University in Torun, 87-100 Toruń, Poland; tgrzyb@cm.umk.pl (T.G.); gorzkiewiczmarta@cm.umk.pl (M.G.)
* Correspondence: el12ab@op.pl (E.Ł.-R.); mgoleb@umk.pl (M.G.)

Abstract: The gut microbiota in patients with food allergy, and the skin microbiota in atopic dermatitis patients differ from those of healthy people. We hypothesize that relationships may exist between gut and skin microbiota in patients with allergies. The aim of this study was to determine the possible relationship between gut and skin microbiota in patients with allergies, hence simultaneous analysis of the two compartments of microbiota was performed in infants with and without allergic symptoms. Fifty-nine infants with food allergy and/or atopic dermatitis and 28 healthy children were enrolled in the study. The skin and gut microbiota were evaluated using 16S rRNA gene amplicon sequencing. No significant differences in the α-diversity of dermal or fecal microbiota were observed between allergic and non-allergic infants; however, a significant relationship was found between bacterial community structure and allergy phenotypes, especially in the fecal samples. Certain clinical conditions were associated with characteristic bacterial taxa in the skin and gut microbiota. Positive correlations were found between skin and fecal samples in the abundance of *Gemella* among allergic infants, and *Lactobacillus* and *Bacteroides* among healthy infants. Although infants with allergies and healthy infants demonstrate microbiota with similar α-diversity, some differences in β-diversity and bacterial species abundance can be seen, which may depend on the phenotype of the allergy. For some organisms, their abundance in skin and feces samples may be correlated, and these correlations might serve as indicators of the host's allergic state.

Keywords: atopic dermatitis; dysbiosis; food allergy; gut; infants; microbiota; skin; 16S rRNA sequencing

Citation: Łoś-Rycharska, E.; Gołębiewski, M.; Sikora, M.; Grzybowski, T.; Gorzkiewicz, M.; Popielarz, M.; Gawryjołek, J.; Krogulska, A. A Combined Analysis of Gut and Skin Microbiota in Infants with Food Allergy and Atopic Dermatitis: A Pilot Study. *Nutrients* **2021**, *13*, 1682. https://doi.org/10.3390/nu13051682

Academic Editor: Carla Mastrorilli

Received: 24 March 2021
Accepted: 12 May 2021
Published: 15 May 2021

Publisher's Note: MDPI stays neutral with regard to jurisdictional claims in published maps and institutional affiliations.

1. Introduction

Although the cause of the growing prevalence of allergic diseases remains unclear, the Old Friends Hypothesis (also called the Biodiversity Hypothesis) has recently been proposed [1]. The hypothesis states that the development of many diseases, including allergy, might be attributed to a lack of exposure to the "right" bacterial strains. Assuming this to be the case, it may be possible that identifying such bacteria could halt the allergic march and foster the development of novel prevention and treatment methods for allergic diseases.

The composition of the human microbiome is not only characteristic of each individual, but also evolves over time [2–7]. In various parts of the child's body, the microbiome begins to form and differentiate within the first six weeks post natum [2,3]. The most pronounced changes in the gut microbiome are usually observed during first two or three years of human life [4,5,8]. Following this, the composition of the gut microbiome remains relatively stable, unless significant environmental changes occur [5]. The early gut microbiome is believed to be shaped by a range of factors, including maternal race-ethnicity, age, diet, antibiotic treatment, marital status, the mode of delivery, health complications after birth, infant feeding method, solid introduction, probiotics intake, exposure to tobacco smoke and pets [2,4,6,7].

Similarly, the skin microbiota evolves over the years according to changes in skin structure and function [9], becoming similar to that of adults by the age of 12–18 months [10] A possible critical window during which microbial-based intervention would be possible is early infancy (i.e., the first six months of life) [10–12]. The composition of skin microbiota is site-specific and, again, depends on a range of factors, including local conditions such as invaginations, pockets, niches and surfaces, as well as moisture, sebum secretion, temperature and exposure to external factors [9,10]. In the neonatal period, the mode of delivery appears to play an important role in shaping the skin microbiome [2]. The microbiome, considered a complex biocenosis where various organisms interact, influences the immune system in a multifaceted way, modulating both innate and adaptive immunity, and shaping allergy development through interactions with the host genome [13–15].

The gut and the skin are crucial barriers through which environmental factors interact with the human body; the two compartments are also massively colonized with distinct microbial communities. Furthermore, they constitute complex immune organs that are fully integrated into the overall immune system. Proper skin and gut functioning is essential for homeostasis, and decreases the risk of allergy. Hence, exposure to the many environmental factors that shape the microbiome of the gut and skin epithelium can alter the epithelial surfaces and thus drive the type 2 immune response, which is known to underlay most, if not all, atopic diseases. Many studies indicate the existence of a gut–skin axis, which is modulated by microbiota and its metabolites [16–18].

Generally, the first symptoms of allergy involve the skin and/or the gastrointestinal tract [19], manifesting as atopic dermatitis (AD) and/or food allergy (FA). These disorders are difficult to diagnose in early infancy, as there are no specific markers, and diagnosis is based mainly on their symptoms. Although previous studies have yielded varying results, it has been found that gut microbiota differed between otherwise healthy subjects and those who were food sensitized (FS) or food allergic (FA) [20,21]. It has been shown that the composition of the microbiota in the neonatal period may be associated with the development of allergies up to the age of one year. In later life, alpha-diversity, beta-diversity, and the richness of individual bacteria are shaped differently in non-allergic and allergic children [3]. It has also been shown that the microbiome structure at three months is associated with the development of sensitization up to one year of age [11]. It also appears that the microbiome of the gut in infancy is a prognostic factor of allergy resolution by eight years of age [12].

The bacterial community structure has also been found to differ between IgE-mediated and non-IgE-mediated FA children [22], and differences can be seen between the microbiota associated with specific food allergies [23–25]. Fieten et al. report no differences in microbial diversity between FA and non-FA children; however, a fecal microbial signature that discriminated between the presence and absence of FA was found in children with AD [26].

Similarly to the gut microbiome, the skin microbiota has also been found to differ between patients with AD and those without [9], and other studies have reported decreased skin microbiome diversity in AD patients [16,27]. The significance of the skin microbiome in AD pathogenesis was confirmed by Laborel-Preneron et al., who showed that *Staphylococcus aureus* colonizing atopic skin promoted inflammation in children aged one to three years, while commensal *S. epidermidis* strains might mitigate this effect [28]. Similarly, Kong et al. found that the abundance of *Staphylococcus* (particularly *S. aureus*) was greater in children with AD, and that microbial diversity increased over the course of treatment [29]. Interestingly, while the development of AD appears to be related to colonization by *S. aureus*, it has also been proven that exacerbations of skin lesions are connected to a reduction in skin microbiome diversity, while improvement is not directly related to a reduction in *S. aureus*, but rather to an increase in bacterial diversity [16]. In addition, prospective studies have identified changes in gut and/or skin microbiota in children, even prior to the development of food sensitivity (FS), FA or AD, with changes being apparent as early as in the third month of life [11,16,30–33]. It has also been proposed that decreased *Staphylococcus* richness in the skin microbiome during early infancy might be a prognostic factor of AD development before 12 months of age, and that colonization by commensal *Staphylococci* might protect against eczema [32].

Current evidence suggests that while the gut microbiota may demonstrate a reduction in microbial diversity before AD onset, no such relationship has been noted for the skin microbiota [16,23]. Furthermore, although gut dysbiosis has been described in AD [23,27,34], it is not clear whether the modulation of gut microbiota can impact skin microbiota and vice versa. The role of the skin microbiome in FA development is also unknown.

To gain a more holistic understanding of how the microbiome of the human body interacts with the immune system, and how it can influence the development of FA and/or AD, it is necessary to study the skin and gut together as part of one combined study. The present study is the first such study to compare the microbiota of two compartments, i.e., gut and skin, in young infants with early onset of FA and/or AD. Indeed, of all the studies regarding the relationship between microbiota and allergy published to date, only one [35] has analyzed multiple compartments.

The present study characterizes the alpha and beta diversity of the fecal, i.e., gut, and skin bacterial communities from infants with FA and/or AD and compares them with those of healthy control subjects.

We hypothesize that:

Hypothesis 1. *Gut and skin alpha diversity differ between children with an allergy (FA and/or AD) and those without;*

Hypothesis 2. *The communities within a single compartment differ according to the clinical status of the child;*

Hypothesis 3. *OTUs characteristic for each compartment-clinical status communities exist;*

Hypothesis 4. *Relationships exist between the gut and skin microbiota, and these differ with regard to clinical status.*

To confirm these hypotheses, V3-V4 16S rRNA gene libraries were generated and sequenced, and the resulting sequences were analyzed bioinformatically.

2. Methods

This is an observational pilot study, constituting part of Stage I of a prospective study.

2.1. Study Group

Participants were recruited from 1 February 2018 until 30 January 2019 among infants hospitalized in the Department of Pediatrics, Allergology and Gastroenterology, Collegium Medicum, Nicolaus Copernicus University, Poland and from Gastrological, Allergological and Nephrological Outpatient Clinics in Bydgoszcz, Poland.

Enrollment was proposed to parents of 245 infants under six months of age. Informed consent was obtained from 203 of them. Out of these patients, 104 did not fulfill the inclusion criteria, or were excluded (Table 1), and the parents of 12 failed to provide the full data needed for the study. The primary cause of disqualification from the study was receipt of known bacterial, viral or fungal infection or antibiotic treatment, either at the present time or in the previous month, due to their impact on the microbiome. The recruitment procedure is summarized in Figure 1.

Table 1. Inclusion and exclusion criteria.

Group	Inclusion Criteria		Exclusion Criteria	
Study group	1.	age: 0–6 months and	1.	immune deficiencies; autoimmunological
	2.	strong suspicion of FA and/or		

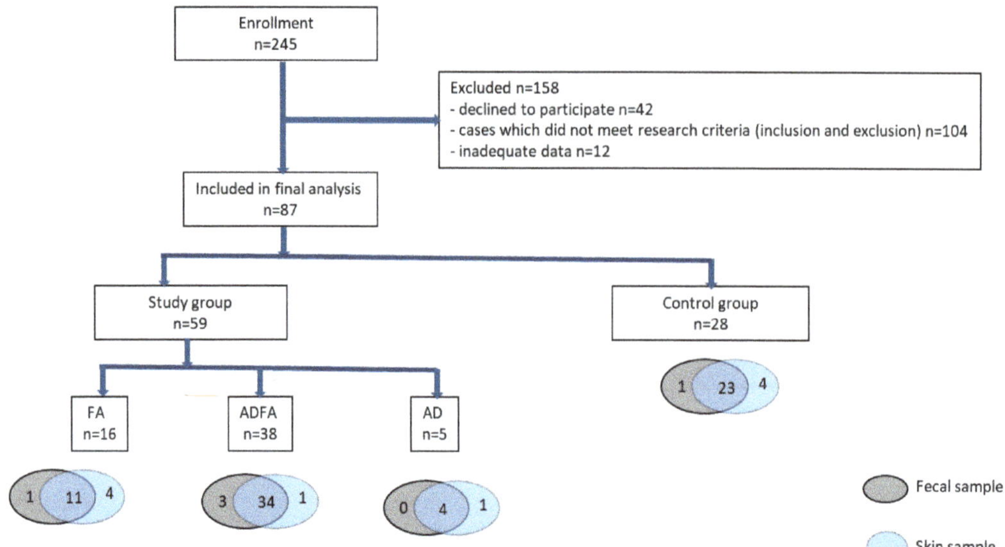

Figure 1. Study design. Flow chart depicting steps involved in patient selection and tests in this study.

Finally, 87 infants were enrolled, 59 of them with suspicion of allergies (henceforth called the "allergic group" or "study group") and 28 healthy infants, comprising the control group. The allergic group comprised 38 patients with symptoms of FA and AD (ADFA group), 16 with symptoms of FA (FA group) and 5 with AD (AD group).

The characteristics of the study and control group are given in Table 2. No significant differences were observed between the FA/AD, FA, AD and control groups in terms of cesarean section, diet (breastfeeding or formula-feeding and solids intake) or family history of allergic diseases. The clinical characteristics of the studied infants are given in Table A1. In the studied group, FA was equivalent to cow's milk allergy (CMA). IgE-sensitization was found in eight (14.8%) infants.

Table 2. Characteristics of the study and control groups.

Parameter	Study Group $n = 59$ (100%)	FA $n = 16$ (100%)	ADFA $n = 38$ (100%)	AD $n = 5$ (100%)	Control Group $n = 28$ (100%)	p
Sex, n (%)						
female	24 (40.7)	6 (37.5)	16 (42.1)	2 (40)	12 (42.9)	>0.05
male	35 (59.3)	10 (62.5)	22 (57.9)	3 (60)	16 (57.1)	
Age at specimen collection (weeks) mean ± SD	15.56 ± 7.00	16.63 ± 6.72	15.92 ± 7.43	16.6 ± 8.39	14.29 ± 6.52	>0.05
Weight at birth (g), mean ± SD	3495 ± 548	3518 ± 493	3571 ± 480	3588 ± 228	3360 ± 686	>0.05
Mode of delivery, n (%)						
Vaginal	43 (72.9)	12 (75)	27 (71.1)	4 (80)	16 (57.1)	>0.05
C-section	16 (27.1)	4 (25)	11 (28.9)	1 (20)	12 (42.9)	
Time of birth, Hbd, mean ± SD	39.3 ± 1.9	39.6 ± 1.7	39.5 ± 1.2	40 ± 0.7	38.7 ± 2.6	>0.05
Apg scale, points, mean ± SD	9.6 ± 1.2	9.6 ± 0.8	9.7 ± 0.6	10 ± 0	9.2 ± 1.9	>0.05
Mode of feeding at entry, n (%)						
Exclusively breastfeeding	28 (47.5)	9 (56.3)	18 (47.4)	1 (20)	13 (46.4)	>0.05
Mixed	7 (11.0)	2 (12.5)	4 (10.5)	1 (20)	3 (10.7)	
Milk formula	24 (40.7)	5 (31.3)	16 (42.1)	3 (60)	12 (42.9)	
Breast-feeding, n (%)						
No, never	8 (13.6)	1 (6.3)	1 (2.6)	1 (20)	5 (17.9)	>0.05
Yes, currently	35 (59.3)	11 (68.8)	12 (57.9)	2 (40)	16 (57.1)	
Solid food intake, Yes, n (%)	11 (18.6)	2 (12.5)	8 (21.1)	1 (20)	8 (28.6)	>0.05
Place of living n (%)						
non-rural	56 (94.9)	15 (93.8)	36 (94.7)	5 (100)	28 (100)	>0.05
rural	3 (5.1)	1 (6.3)	2 (5.3)	0	0	
Average living space per person, (m^2) mean ± SD	24.6 ± 14.1	27.4 ± 17.9	24.4 ± 13.1	29.2 ± 19.3	22.4 ± 12.3	>0.05
Comiss score, points mean ± SD	12.71 ± 3.2	12.73 ± 1.54	13.62 ± 1.96	6 ± 0.65	5.63 ± 2.74	<0.00001
Parent's age, years						
Mother	28.9 ± 4.6	29 ± 4.7	30.1 ± 3.9	29 ± 5.1	27.29 ± 4.9	>0.05
Father	31.7 ± 5.2	32 ± 5.4	32.37 ± 4.7	31.2 ± 4.1	30.6 ± 5.9	
Parent's education, n (%)						
Mother Basic	6 (10.2)	4 (25)	1 (2.6)	1 (20)	6 (21.4)	0.03
Secondary	19 (32.2)	4 (25)	12 (31.6)	3 (60)	12 (42.9)	
Higher	34 (57.6)	8 (50)	25 (65.8)	1 (20)	10 (35.7)	
Father Basic	10 (16.9)	5 (31.3)	5 (13.2)	0	7 (25)	>0.05
Secondary	23 (39)	5 (31.3)	15 (39.5)	3 (60)	14 (50)	
Higher	26 (44.1)	6 (37.5)	18 (47.4)	2 (40)	7 (25)	
Atopy in family, n (%)						
Yes	43 (72.9)	13 (81.3)	28 (73.7)	2 (40)	15 (53.6)	>0.05
Mother	29 (49.2)	10 (62.5)	18 (47.4)	1 (20)	6 (21.4)	0.025
Father	19 (32.2)	4 (25)	14 (36.8)	1 (20)	6 (21.4)	>0.05
Siblings	14 (56.0)	4 (57.1)	10 (62.5)	0	4 (40)	>0.05
Siblings, n (%)	25 (42.4)	7 (43.8)	16 (42.1)	2 (40)	10 (35.7)	>0.05
Pets at home, n (%)						
Yes	25 (42.4)	8 (50)	15 (39.5)	2 (40)	13 (46.4)	>0.05
Dog	20 (33.9)	6 (37.5)	12 (31.6)	2 (40)	9 (32.1)	>0.05
Cat	5 (8.5)	4 (25)	1 (2.6)	0	3 (10.7)	>0.05
Tobaco smoke exposure, n (%)						
Mother active during pregnancy	2 (3.4)	1 (6.3)	1 (2.6)	0	3 (10.7)	>0.05
Mother during pregnancy passive	12 (20.3)	5 (31.3)	7 (18.4)	0	8 (28.6)	>0.05
Mother during lactation	1 (1.7)	0	1 (2.6)	0	3 (10.7)	>0.05
Child passive	14 (23.7)	5 (31.3)	7 (18.4)	2 (40)	10 (35.7)	>0.05
Antibiotics during pregnancy (3rd tr), n (%)	4 (6.8)	0	4 (10.5)	0	4 (14.3)	>0.05

p-value of Fisher's exact test is given.

2.2. Preliminary Questionnaire

The participants completed the modified ISAAC questionnaire [36]. It comprised questions on the family history of atopic disease, as well as the presence of symptoms in the child and their relationship with diet, perinatal conditions, social background and demographic data.

2.3. FA Diagnosis

The diagnosis of FA was based on history, physical examination, COMISS evaluation, specific IgE in the blood, elimination and oral provocation tests.

COMISS (Cow's milk-related symptoms score) was assessed based on the standard questionnaire [37]. Scores ≥ 12 were regarded as indicating a high probability of FA. Other factors taken into consideration were: (i) in breastfed children: remission on elimination diet and relapse on reintroduction of milk, according to EAACI [38]; in children fed with milk formula: remission on elimination diet and positive outcome of open oral food provocation test according to ESPGHAN and EAACI [19,39], (ii) typical symptoms related to cow's milk protein intake (frequent regurgitation or vomiting; extended periods of diarrhea with negative microbiological tests; blood in stools; chronic poor weight gain on milk inclusion; iron deficiency anemia due to cryptic or macroscopic blood loss with stools and not due to infection or insufficient iron intake); eosinophilic enteropathy confirmed with endoscopy; running nose, wheezing and chronic coughing unrelated to infection; atopic eczema (moderate or severe, defined as persistent or frequently recurring eczema with typical morphology and distribution, requiring frequent need for prescription topical corticosteroids, calcineurin inhibitors, despite appropriate use of emollients.); urticarial (i.e., unrelated to infections, drug intake and other causes) swelling of skin or hives; recurrent itchy or flushed skin; anaphylaxis [19].

2.4. Atopic Dermatitis (AD) Diagnosis

Atopic dermatitis (AD) was diagnosed based on the UK Working Party's criteria with modifications necessary for infants [40,41], i.e., itchy skin (or parental report of scratching or rubbing) and at least three of the following symptoms:

- history of skin rash affecting the flexures (folds of elbows, behind knees, fronts of ankles), cheeks, neck, around eyes or outer surfaces of limbs;
- history of atopic diseases in a first-degree relative;
- history of generally dry skin;
- visible eczema involving cheeks, forehead, flexures and extensor/outer surfaces of limbs.

AD severity was assessed based on SCORing Atopic Dermatitis (SCORAD) scale at study entry [42]. AD patients were further stratified into two groups based on the presence of accompanying FA symptoms:

- AD without FA;
- AD with FA, i.e., eczema despite emollients; eczema and other typical symptoms of FA; significant improvement after elimination diet and deterioration after reintroduction of presumed allergen into the diet.

2.5. Sample Collection

Fecal and skin samples were collected once, during the visit at the time of study enrollment. Sterile, disposable equipment was used throughout the sampling procedure. Freshly evacuated feces in diapers were gathered into collection tubes and frozen for transport to the laboratory. Skin swabs were collected from non-lesional, healthy looking skin on the cheeks; in cases where the cheeks were atopic, sampling was performed from any other healthy skin region excluding the diaper area, usually from the dorsal surface of the forearm. All of the samples were stored at $-80\ °C$ until further processing.

2.6. Determination of sIgE

Serum IgE was quantified with Polycheck test (BioCheck GmbH, Leipzig, Germany) according to the manufacturer's protocol. The detection limit was assessed as 0.35 kU/l. The patient was regarded as sensitized if specific IgE (sIgE) concentration exceeded the detection limit. Main food allergens (cow's milk, hen's egg, peanuts, soybean, wheat, hazelnut, codfish) and inhalant ones (house dust mites *Dermatophagoides pteronyssinus* and *D. farinae*, cat dander, dog dander, birch, hazel, and grass pollen, olive and Alternaria spores) were tested.

2.7. Metagenomic Analysis

DNA was isolated from clinical samples using bead-beating (on PowerLyzer, Mo-Bio) combined with flocculation and silica columns. V3-V4 bacterial 16S rRNA gene fragments were amplified and converted into Illumina sequencing libraries as previously described [43]. The libraries were sequenced on MiSeq using 600 cycles v.3 kit (Illumina) at CMIT NCU.

2.8. Bioinformatics and Statistical Analyses

The sequencing reads were processed as described earlier [43]. Briefly, the reads were denoised, merged and checked for the presence of chimaeras in dada2 [44]; they were then classified using SILVA v.132 reference [45]. Following this, OTUs were constructed, and a shared OTU table was calculated with Mothur [46]. Ecological analyses were performed in R using functions implemented in the vegan [47] and GuniFrac packages [48,49]. An unweighted UniFrac distance matrix was generated using the GuniFrac function from the rarefied OTU table and tree using the Relaxed Neighbor Joining algorithm implemented in clearcut [50]. Unconstrained ordination was performed using non-metric multidimensional scaling (NMDS), implemented in the metaMDS function of vegan. The grouping significance was tested with PERMANOVA (adonis function) using 999 permutations; dbRDA was performed on the unweighted UniFrac distance matrix using dbrda. The significance of the dbRDA model was tested with anova.rda using 999 permutations. p-value < 0.05 was considered significant.

Differences in taxa abundance were assessed with ANOVA (aov function in R); the normality of data was checked with the Shapiro–Wilk test (shapiro.test) and homogeneity of variance by Levene's test (levene.test of the lawstat packege). Where the assumptions were not fulfilled, the Kruskal–Wallis test was used (kruskal.test). p-values were corrected for multiple comparisons using Benjamini-Hochberg FDR (p.adjust); a significance level of 0.05 was used. Clinical categorical data were analyzed using two-tailed Fisher's exact test in R (fisher.test); again, a significance level of 0.05 was used.

2.9. Additional Information

All methods were carried out in accordance with relevant guidelines and regulations. The study protocol was approved by the local Ethics Committees of the Institutional Review Board of CM Bydgoszcz NCU Torun, Poland (765/2017). Informed written consent was obtained from the parents of all participants prior to enrollment.

2.10. Data Availability

Sequence data generated during this project are available in the SRA database under BioProject no. PRJNA657878.

3. Results

3.1. No Significant Differences in Alpha Diversity Indices

No significant differences in diversity (Shannon's H'), species richness (observed number of OTUs) and evenness (Shannon's E) were found between controls and the participants with either AD, ADFA or FA, regardless of the compartment (skin and gut). Similarly, no differences were found between the group of allergic patients and those without an allergy. However, the diversity of skin microbiota tended to be lower in AD patients (Figure 2). No correlations of any alpha-diversity indices were observed between the skin and feces, regardless of the analyzed group.

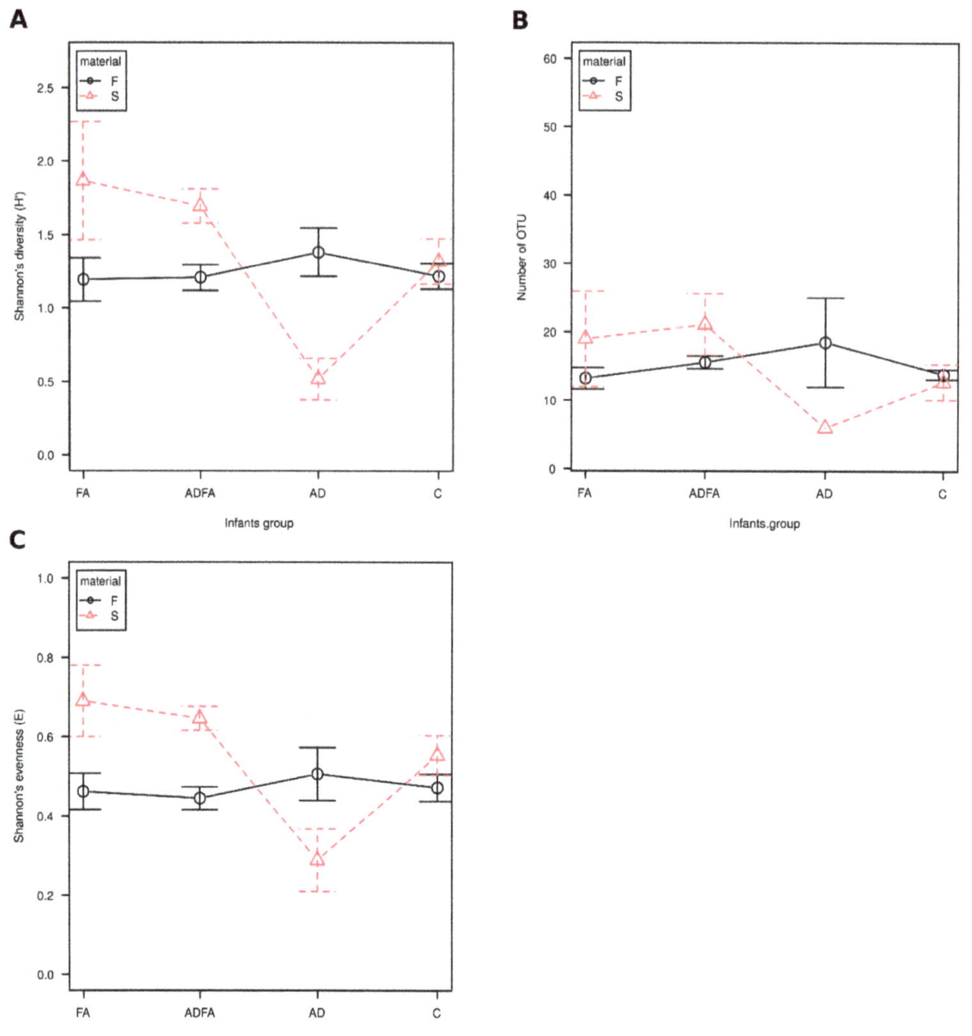

Figure 2. α-diversity of feces and skin microbiota in infants with FA, AD and ADFA. F—feces, S—skin, FA—food allergy, AD—atopic dermatitis, ADFA—atopis dermatitis and food allergy, C—control group. (**A**) Shannon's diversity index (H′), (**B**) Observed number of OTUs, (**C**) Shannon's evenness (E).

3.2. Gut and Skin Microbiota Differ According to Clinical Status

Significant differences were observed between the control, AD, ADFA and FA groups regarding fecal sample community structure (unweighted UniFrac distance, dbRDA, $p = 0.003$; Figure 3A). In addition, differences were found between ADFA and FA ($p = 0.04$) and between ADFA and control ($p = 0.05$).

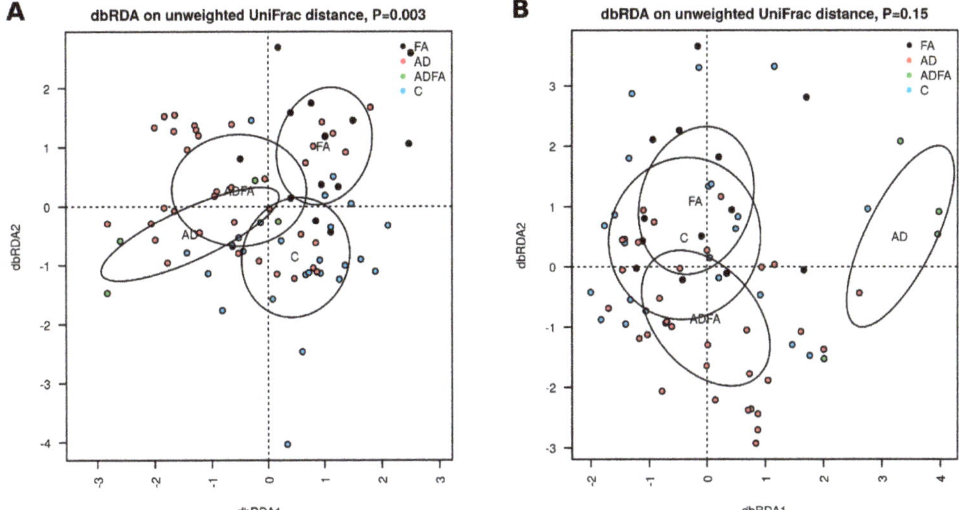

Figure 3. (**A,B**). β-diversity of fecal and skin microbiota in the studied groups. β-diversity: distance based redundancy analysis (db RDA) on unweighted distance metrices calculated on rarified community data for: A—feces, B—skin, 95% confidence elapses for show significance of grouping according to clinical status was tested FA—food allergy group, AD—atopic dermatitis group, ADFA—atopic dermatitis and food allergy group, C—control group.

Although no significant overall differences in skin communities were observed between groups ($p = 0.15$), significant differences became apparent when age (treated as continuous variable) was added to the overall model ($p = 0.013$). Regarding individual group comparisons, significant differences in the skin microbiome were found between AD and FA ($p = 0.014$), AD and control ($p = 0.01$), as well as AD and ADFA ($p = 0.02$) (Figure 3B).

3.3. There Are Taxa Characteristic for Compartments and Clinical Status

Any taxa whose abundance differed significantly between groups were regarded as characteristic for the group where the given taxon was more highly represented (Figure 4).

The members of the Xanthomonadales and Xanthomonadaceae were more abundant in the skin communities of all allergic (AD + ADFA + FA) patients, than in healthy children ($p = 0.04$ and $p = 0.004$, respectively); however, the reverse was true for the order Lactobacillales ($p = 0.045$), family Streptococcaceae ($p = 0.045$), genus *Streptococcus* ($p = 0.044$), and OTU13 belonging to genus *Bradyrhizobium* ($p = 0.019$).

In the gut of allergic children, the following taxa were more abundant: phylum Bacteroidetes ($p = 0.03$), class Bacteroidia ($p = 0.03$), order Bacteroidales ($p = 0.046$), family Bacteroidaceae ($p = 0.043$) and genus *Bacteroides* ($p = 0.043$). At the OTU level, OTU35, affiliated with *Parabacteroides*, and OTU21, belonging to *Bacteroides*, were more abundant in allergic children ($p = 0.039$ and $p = 0.008$, respectively), while OTU94 (*Fusicatenibacter saccharivorans*; $p = 0.049$), OTU68 (*Lactococcus lactis*; $p = 0.035$) and OTU30 (*Serratia marcescens*; $p = 0.035$) were more common in the control group.

Comparing all four groups allowed the identification of taxa specific for a particular clinical picture. The highest abundance of phylum Proteobacteria was found in skin samples of AD patients; lower levels were observed in ADFA and the lowest in the FA and control groups ($p = 0.02$). Sequences affiliated with Streptococcaceae were most frequent in libraries derived from skin samples of healthy children, but were absent from patients with skin symptoms (AD and ADFA; $p = 0.04$), the same applied to genus *Streptococcus*

($p = 0.039$). At the OTU level, only low abundance sequences (less than 10 reads across all groups) were found to be characteristic of a particular clinical picture.

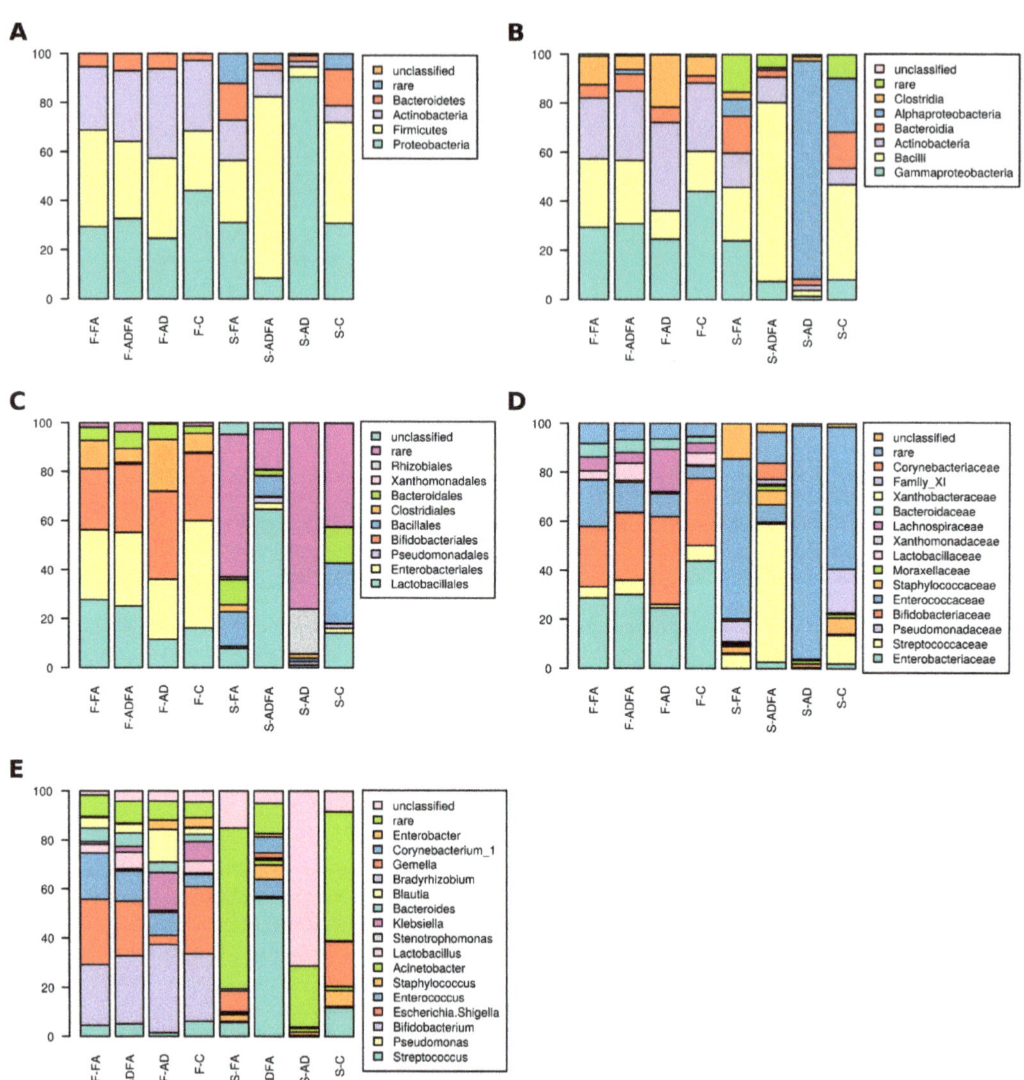

Figure 4. Taxonomic composition: (**A**). phyla level, (**B**). class level, (**C**) order level, (**D**). family level, (**E**). genus level in feces and skin microbiota of the studied children. F-FA—feces in food allergy group, F-AD—feces in atopic dermatitis group, F-ADFA—feces in atopic dermatitis and food allergy group, F-C—feces in control group, S-FA–skin in food allergy group, S-AD—skin in atopic dermatitis group, S-ADFA—skin in atopic dermatitis and food allergy group, S-C—skin in control group.

In the feces samples, phylum Bacteroidetes was more abundant in AD children, less frequent in ADFA and absent from healthy controls ($p = 0.043$); the same was observed to orders Bacteroidales and Xanthomonadales ($p = 0.038$ and $p = 0.03$, respectively), families Xanthomonadaceae and Bacteroidaceae ($p = 0.03$ and $p = 0.043$, respectively) and genera *Stenotrophomonas* and *Bacteroides* ($p = 0.03$ and $p = 0.041$, respectively). In contrast, the abundance of genus *Enterobacter* was higher in children with no intestinal symptoms (AD and control) but lower in those with a food allergy (FA and ADFA; $p = 0.009$). Differences in OTU abundance (Table 3) were also observed.

Table 3. Differentially abundant OTUs.

	OTU	Abundance in Different Groups	p
11	Stenotrophomonas maltophilia	AD > ADFA > FA, C	0.030
21	Bacteroides	FA, ADFA > AD > C	0.030
35	Parabacteroides	AD, ADFA > FA, C	0.010
62	Rhizobium (Agrobacterium fabrum)	C > AD > FA, ADFA	0.023
85	Veilonella dispar	FA > ADFA > AD, C	0.015

p-value of Kruskal–Wallis test is given.

To identify the taxa potentially involved in AD and FA pathogenesis, the results for the control group were compared with those from all patients with symptoms either on the skin (AD + ADFA) or in the gut (FA + ADFA) (Table 4). Libraries derived from skin of AD + ADFA patients displayed higher levels of the members of the order Xanthomonadales ($p = 0.029$) and the families Xanthomonadaceae and Corynebacteriaceae ($p = 0.029$ and $p = 0.036$, respectively). In addition, in the feces-derived libraries, class Bacteroidia ($p = 0.007$), order Bacteroidales ($p = 0.004$) and family Bacteroidaceae were found to be more common in AD + ADFA than in healthy children.

Table 4. Differentially represented OTUs—characteristic either for AD or FA.

Group		Skin				Feces		
		OTU	p	More Abundant in		OTU	p	More Abundant in
AD + ADFA vs C	87	Acinetobacter variabilis	0.012	AD + ADFA	11	Stenotrophomonas maltophilia	0.012	AD + ADFA
					35	Parabacteroides	0.006	
	2	Streptococcus sp.	0.044	C	41	Bacteroides dorei	0.043	
					21	Bacteroides	0.008	FA + ADFA
FA + ADFA vs C	13	Bradyrhizobium sp	0.027	C	30	Serratia marcescens	0.042	C
					68	Lactococcus lactis	0.042	

p-value of Kruskal–Wallis test is given.

Libraries from the skin samples of children with FA (FA + ADFA) were more abundant in bacteria from the order Xanthomonadales ($p = 0.016$) and family Xanthomonadaceae ($p = 0.016$) than those from the feces samples. Differences at the OTU level were also observed: in the skin samples, OTU2 and OTU13 were more abundant among control patients than the AD and FA groups, respectively. In the feces samples, OTUs 11, 35 and 41 were more abundant in the AD group than controls, while only OTU21 was more abundant in FA patients than the controls (Table 4).

3.4. There Are Taxa Whose Abundance was Positively Correlated in Feces and on Skin

Positive correlations were observed between the skin and gut libraries for the genus *Enterobacter*, regardless of patient clinical status ($\rho = 0.26$, $p = 0.04$). Similarly, positive correlations were observed for the genus *Gemella* for all allergic children (AD + ADFA + FA) ($\rho = 0.45$, $p = 0.02$), and for the genera *Lactobacillus* and *Bacteroides* in the control group ($\rho = 0.48$, $p = 0.03$ and $\rho = 0.49$, $p = 0.02$, respectively). No negative correlations were observed between the skin and feces libraries.

The abundance of 13 OTUs belonging to 11 genera was positively correlated between respective skin and feces samples (Table 5). Among them, OTU2 (*Streptococcus*) and OTU31 (*Lactobacillus*, nearest species *L. gasseri*) were correlated in both the allergic and control groups. Interestingly, no negative correlations were observed. In allergic subjects, the correlated OTUs were affiliated with Gram-positive phyla (Firmicutes and Actinobacteria), while those correlated in healthy subjects included the Gram-negative taxa *Haemophilus* (Gammaproteobacteria) and *Bacteroides* (Bacteroidetes).

Table 5. OTUs whose abundance on skin and in feces is correlated.

Infants with Allergy Symptoms				Healthy Infants			
	OTU	rho	*p*		OTU	rho	*p*
2	Streptococcus sp.	0.34	0.030	2	Streptococcus sp.	0.17	0.030
14	Gemella sp.	0.45	0.002	25	Acinetobacter sp	0.72	0.0002
16	Bifidobacterium scardovii	0.35	0.020	31	Lactobacillus gasseri	0.64	0.0002
17	Corynebacterium nuruki	0.72	0	32	Haemophilus haemolyticus	0.55	0.010
20	Rothia mucillaginosa	0.45	0.002	40	Bacteroides ovatus	0.61	0.004
31	Lactobacillus gasseri	0.35	0.020	53	Schaalia odontolytica	0.46	0.030
38	Lactobacillus sp.	0.68	0	74	Actinomyces graevenitzi	0.72	0.0002
92	Lactobacillus salivarius	1	0				

Spearman's rho and *p*-value are given.

4. Discussion

We examined the diversity and community structure of both skin and gut (feces) microbiota in young infants with allergies symptoms, using 16S rRNA gene amplicon sequencing. In addition to being the first of its type to be performed on Polish patients, the study has four key strengths. Firstly, the gut and skin microbiota were examined simultaneously and compared. Secondly, a novel analysis was performed on the skin microbiota in infants with FA. Thirdly, the study group was characterized by a range of clinical pictures, which allowed the differences in the microbiotas to be highlighted. Finally, unlike most previous studies, the participants were under six months of age; this allows the relationship between the microbiota and the development of allergy to be assessed in an early stage of life.

The bacterial communities found in human gut and skin vary considerably between individuals and with regard to developmental stages, as do views on their relationship with allergy development [13,15,16,22,26]. These differences might stem from the profile of the studied group, e.g., the number of patients, clinical picture, demographics and environmental conditions, as well as on the methodology used, such as the primer system, PCR conditions, sequencing technology and the database used for classification [51]. This variability might explain why our results disagree with many published studies.

However, similarly to other studies [3,23,26,32,52–54], no significant differences in bacterial alpha diversity were found between allergic and non-allergic infants, regardless of the type of allergy or compartment, i.e., skin or feces. These results suggest that, contrary to results obtained for older patients [29,55], decreased diversity does not appear to be related to allergy symptoms, even in the case of AD. This might be due to the fact that the microbial community is still developing in young infants; nevertheless, this observation indicates that neither species richness nor diversity can be used as predictors of allergy in infants under one year of age. However, in contrast to our present findings, a number of previous studies have reported significant variation in gut microbiota α-diversity between infants with FA and healthy children [25,56]. It is possible that this variation could be attributed to the differences between the groups of participants: age (10 months [25] vs four months in our study), feeding method (currently breastfeeding 39% [25] vs 59% in our study), type of allergy (IgE-mediated [25] vs IgE + nonIgE-mediated (56 and our study)), type of allergenic food (egg [25] vs milk in our study) or, finally, genetic and environmental difference (China [56] vs Europe in our study).

We observed a significant correlation between fecal bacterial community structure and allergy type, even though the communities varied greatly within one group. This result is similar to those obtained previously [3,12]. Interestingly, both skin and fecal microbial communities differed significantly between the AD and ADFA groups, suggesting that these two clinical conditions differ in their immunological background; this difference is known to influence the microbiome, and vice versa [13,15]. Skin communities also differed, suggesting that assessing skin swabs could be a diagnostic tool for particular phenotypes of allergy. It could be possible to devise a classifier (e.g. using RandomForest framework) [57] to predict the type of allergy based on community structure. Such a tool could be useful in clinical practice, aiding in diagnosis. However, this will require further studies based on larger numbers of patients, i.e., providing greater statistical power.

Some of the taxa whose abundance was found to differ between allergic and healthy infants have previously been found to be related to allergies, although the direction of these relations varied. We found that the abundance of Bacteroidales and their member taxa, down to the level of genus (*Bacteroides* and *Parabacteroides*) in the libraries derived from feces, was related to certain types of allergy. This relationship might be driven by the production of the epithelial barrier-breaching agent phenol/*p*-cresol, which is known to be produced by the members of the Bacteroidaceae [58]. Previous studies have noted an increase in Bacteroidales members correlated with various types of allergy [53,59,60]. On the other hand, contrary to our present findings, a number of studies have reported a higher abundance of Bacteroidetes in healthy individuals [11,20,56]; however, these studies employed groups comprised of participants from different countries (Canada, Taiwan, China) and at different ages (thirteen months [20] vs three months, twelve months [11], three months [56], four months in our study).

Bacteria of the order Xanthomodales, particularly those of the genus *Xanthomonas*, are frequently found on healthy skin, albeit at a low abundance [61]. In the present study, they were associated with skin communities from allergic patients: the first such report. The abundance of *Xanthomonas* was found to increase after successful treatment of AD with emolients [62]. Although the precise role of these bacteria in skin dysbiosis remains unclear, it is possible that their keratinolytic properties may be involved. Alternatively, they might participate in shaping the microbiome structure via microbe–microbe interactions.

Many authors have indicated that Staphylococcus plays an important role in the relationship between AD and skin colonization [29,32,63]. Kong et al. reported an increase in the number of Staphylococcus, including both *S. aureus* and *S. epidermidis*, in patients aged 2–15 years with AD exacerbation, and that topical skin treatments increased the proportion of Streptococcus, Propionibacterium and Corynebacterium [29]. Similarly, our findings indicated that the children with AD were characterized by a smaller proportion of Streptococcus than the controls.

Assuming the existence of the gut–skin axis in humans, i.e., changes in the skin microbiome that have been found to modulate the gut microbiome [64], and changes in skin physiology that appear to induce changes in the gut microbiota [65], we hypothesize that microbiota of the skin would influence that of the gut and vice-versa. This relationship was examined in the present study.

It seems likely that the gut microbiome influences the health status of skin [66–69], particularly in the pathogenesis of AD [17]. Certain organisms, such as *Escherichia*/*Shigella* or *Veillonella*, were found to be more abundant in the feces of AD patients, while others (*Streptococcus, Haemophilus*) were more prominent in those of healthy subjects [69].

Our observations suggest that gut dysbiosis might be involved in the development of atopic skin inflammation, even though different organisms were differentially abundant in feces and skin libraries. Similar observations have been reported previously [23,33,35,54,66,69].

In spite of the great variability in bacterial communities observed in our samples, we found certain bacterial taxa to be characteristic of particular clinical conditions, and identified a positive correlation between the abundance of certain taxa (*Gemella, Lactobacillus,*

Bacteroides) in the skin samples with those in the fecal samples. This observation, made for the first time, supports the skin–gut axis hypothesis [16–18,64,66].

Certain taxa found to be correlated between compartments in the present study have previously been noted as having an influence on allergy, e.g., *Bacteroidetes* [53] and lower taxa, particularly the genera *Bacteroides* and *Parabacteroides* in the intestines [60,70] or the genus *Staphylococcus*, particularly *S. aureus*, in the skin [29,32,63]. The taxa that correlated between compartments differed depending on the clinical picture, which could be explained by two factors: (i) transmission of microorganisms from the skin to the guts and (ii) selection by that immune system. Indeed, Beijerinck notes, "everything is everywhere, but the environment selects", or rather, microbial communities are shaped by a combination of environmental filtering and stochasticity [71,72]. Therefore, our observed correlation indicates that selection operated similarly in the compartments in question.

Hence, we hypothesize that the key factor affecting the community assembly of the skin and feces is the immunological profile of the host. If the organisms co-vary, then the immune system exerts similar pressure on them in different compartments, and their abundance depends on the initial load. As the sets of strains differ between healthy and allergic children, it is possible that the two groups demonstrate different immune responses. Such correlations may well be of practical importance and serve as indicators of the patient's state of health. Interestingly, although the correlated organisms did not predominate, they were nevertheless found to be involved in allergy pathogenesis [73–75].

Our findings indicate that the abundance of the genus *Gemella* was positively correlated between the feces and skin communities in all allergic children, but not in healthy ones. No similar observation was found in the literature, but some relationships between *Gemella* and allergic inflammation seems to be possible. Taylor et al. [73] report a positive correlation between *Gemella* and sputum eosinophil percentage in patients with asthma.

One of the OTUs whose abundance was correlated between the gut and the skin was *Bifidobacterium scardovii*; however, this was only observed in the allergic group. This could be explained by the fact that although most of the bacteria belonging to the genus *Bifidobacterium* have been recognized as beneficial to human health, *Bifidibacterium scardovii* is the one of five strains of the *Bifidobacterium* which are not—its presence has been associated with clinical conditions such as urinary infections [76].

Among the other OTUs with a positive correlation between fecal and skin microbiota in the allergic patients was *Corynebacterium nuruki*. Generally, *Corynebacteria* are believed to have a rather positive impact on human health [77–79]; however, *Corynebacterium kroppenstedtiian* has been associated with the microbiome of lesioned skin in AD patients [75] and the abundance of the genus *Corynebacterium* has been associated with chronic rhinosinusitis [74]. A novel finding in the present study was that OTU20, identified as *Rothia mucillaginosa*, also demonstrated a positive correlation between the skin and feces of allergic patients. Surprisingly, a positive correlation was also demonstrated for OTU 38 and 92, belonging to the genus *Lactobacillus*, which is known to bestow a protective effect against allergy [80].

Despite its strengths, our study has also some limitations. Most importantly, the group was quite small. This is particularly the case for the AD group, which can be attributed to the selection bias associated with recruiting the patients in an academic center specialized in FA management. In addition, despite the wealth of experience of the participating center, FA remains difficult to diagnose in such young infants; therefore, continuation of the study is planned to confirm our findings by diagnosing the children when they are older. The study group was also heterogeneous with respect to certain important characteristics, such as breastfeeding, introduction of solid food or probiotic intake; however, no statistically significant differences were found between the study and control groups with regard to these variables. It was also heterogeneous with respect to the severity of symptoms at the time of sample collection. Due to the non-specificity and diversity of allergy symptoms in such young infants, it would be very difficult to collect a group with identical clinical parameters.

Another weakness of the study lies in the selection of primers; they could be not well-suited for both compartments, and some organisms, such as *Staphylococcus*, might not be well classified. Skin microbiota is usually assessed based on the V1-3 region and it is not encouraged to sequence V4 only [81]. On the other hand, our system generated a long fragment (450 bp), allowing for classification down to the species level for most organisms. It was also found that the V3-V4 region yielded better results in terms of diversity recovered than the V1-V2 region [82]. Finally, we collected swabs from healthy skin because not all children were in the acute phase of AD at the time of enrollment; however, we think that this approach was justified as Leung et al. report significant differences between the microbiota of non-lesioned, apparently healthy skin of children with AD and ADFA, but no such difference was observed for tissue displaying a skin rash [63].

Understanding the interrelationship between the skin and the gut may allow the design of new treatment models for skin diseases based on the manipulation of the gut microbiota, and conversely, for gastrointestinal diseases through the manipulation of skin microbiota. Moreover, further research on the microbiome may allow the determination of specific patterns of bacteria in multiple body compartments, which are prognostic factors for the development of allergies, and various clinical symptoms (AD, FA). They can also lead to the selection of specific probiotics for treating allergic conditions, either applied to the skin or applied enterally; our analysis suggests that *Acinetobacter* spp., *Haemophilus haemolyticus*, *Bacteroides ovatus*, *Schaalia odontolytica* and *Actinomyces graevenitzi* may be potential candidate bacteria for such treatment, their numbers being correlated with the feces and skin of healthy children.

5. Conclusions

Although the gut and skin microbiota of infants with early symptoms of allergy demonstrate similar α-diversity to those of healthy infants, some differences in β-diversity and the abundance of some bacteria can be seen, which may depend on the phenotype of the allergy. Moreover, the gut and the skin microbiota appear to be related to one another; however, this relationship may differ between patients with allergies and those without. These correlations might serve as indicators of the allergic state of the host.

Author Contributions: E.Ł.-R.–collected clinical data, interpreted data, wrote the first draft of the manuscript, participated in study design; M.G. (Marcin Gołębiewski)—participated in study design, performed bioinformatics and statistical analyses, interpreted data, participated in writing; M.S.—performed experiments; T.G.—provided funding and supervised the study, M.G. (Marta Gorzkiewicz)–contributed to the final version of the manuscript, M.P.—collected clinical data, J.G.— collected clinical data, A.K.—participated in study design, collected clinical data, interpreted data, participated in writing. All the authors approved the final manuscript as submitted and agree to be accountable for all aspects of the work. All authors have read and agreed to the published version of the manuscript.

Funding: This study was financed through grant from National Science Centre, Poland, award number 2017/25/B/NZ5/00141 to TG. The funder had no role in study design, analyzing data and writing the manuscript.

Institutional Review Board Statement: The study was conducted according to the guidelines of the Declaration of Helsinki, and approved by the Bioethics Committee at Nicolaus Copernicus U niversity in Torun, Collegium Medicum in Bydgoszcz, protocol code KB765/2017, approved 12 December 2017.

Informed Consent Statement: Informed consent was obtained from all subjects involved in the study.

Data Availability Statement: Sequence data generated during this project are available in the SRA database under BioProject no. PRJNA657878.

Conflicts of Interest: The authors declare no conflict of interest.

Appendix A

Table A1. Supl. Clinical characteristic of the study group.

Parameter	Study Group $n = 59$ (100%)	FA $n = 16$ (100%)	ADFA $n = 38$ (100%)	AD $n = 5$ (100%)
Clinical symptoms, n (%)				
skin				
rush, erythema, xerosis	46 (77.96)	3 (18.75)	38 (100)	5 (100)
itching	0	0	17 (44.73)	2 (40)
urticarial/angiooedema	5 (8.47)	1 (6.25)	4 (10.25)	0
gastrointestinal				
colic/abdominal pain	43 (72.88)	12 (75)	29 (76.31)	2 (40)
vomiting/massive regurgitation	26 (44.06)	6 (37.5)	19 (50)	1 (20)
weight loss/poor weight gain	9 (15.25)	0	9 (23.68)	0
loose of apetite /flatulence/reflection	6 (10.16)	2 (12.5)	4 (10.52)	0
diarrhea	24 (40.60)	8 (50)	16 (42.10)	0
blood/mucus in stool	23 (38.98)	9 (56.25)	13 (34.21)	1 (20)
constipation	2 (3.38)	1 (6.25)	0	1 (20)
others				
runny nose/snuffles	9 (15.25)	5 (31.25)	4 (10.52)	0
wheezing	4 (6.77)	1 (6.25)	3 (7.89)	0
recurrent cough	2 (3.38)	1 (6.25)	1 (2.63)	0
pallor/cyanosis after ingestion	6 (10.16)	3 (18.75)	3 (7.89)	0
sweating, weakness after eating food	3 (5.08)	2 (12.5)	1 (2.63)	0
sIgE, n (%)				
at least >1	10 (16.94)	2 (12.5)	8 (21.05)	0
cow's milk	0	2 (12.5)	6 (15.78)	0
egg	0	0	1 (2.63)	0
wheat	0	0	1 (2.63)	0
peanut	0	0	1 (2.63)	0
SCORAD index, points				
mean ± SD	23.77 ± 19	-	23.03 ± 17.4	29.4 ± 20.8
Atopic dermatitis severity, n (%)				
Mild <20 pkt	22 (37.28)	0	20 (52.63)	2 (40)
Moderate 20–40 pkt	14 (23.72)	0	13 (34.21)	1 (20)
Severe >40 pkt	7 (11.86)	0	5 (13.15)	2 (40)
AD onset (<12 weeks), n (%)	35 (59.32)	-	31 (81.57)	4 (80)
mean ± SD	6.4 ± 5.4	-	6.3 ± 5.3	6.6 ± 5.5
Types of CMA from gastrointestinal tract				
FPIP	19 (32.20)	8 (50)	11 (28.94)	0
FPIES	6 (10.16)	2 (12.5)	4 (10.52)	0
FGIDs	21 (35.59)	5 (31.25)	16 (42.10)	0
IgE-mediated FA	8 (13.56)	2 (12.5)	6 (15.78)	0

References

1. Rook, G.A.; Lowry, C.A.; Raison, C.L. Microbial 'Old Friends', immunoregulation and stress resilience. *Evol. Med. Public Health* **2013**, *1*, 46–64. [CrossRef]
2. Chu, D.M.; Ma, J.; Prince, A.L.; Antony, K.M.; Seferovic, M.D.; Aagaard, K.M. Maturation of the infant microbiome community structure and function across multiple body sites and in relation to mode of delivery. *Nat. Med.* **2017**, *23*, 314–326. [CrossRef]
3. Shen, X.; Wang, M.; Zhang, X.; He, M.; Li, M.; Cheng, G.; Wan, C.; He, F. Dynamic construction of gut microbiota may influence allergic diseases of infants in southwest China. *BMC Microbiol.* **2019**, *19*, 123. [CrossRef] [PubMed]
4. Jungles, K.N.; Jungles, K.M.; Greenfield, L.; Mahdavinia, M. The infant microbiome and its impact on development of food allergy. *Immunol. Allergy Clin. N. Am.* **2021**, *41*, 285–299. [CrossRef] [PubMed]
5. Gensollen, T.; Iyer, S.S.; Kasper, D.L.; Blumberg, R.L. How colonization by microbiota in early life shapes the immune system. *Science* **2016**, *352*, 539–544. [CrossRef] [PubMed]
6. Obermajer, T.; Grabnar, I.; Benedik, E.; Tušar, T.; Robič Pikel, T.; Fidler Mis, N.; Bogovič Matijašić, B.; Rogelj, I. Microbes in infant gut development: Placing abundance within environmental, clinical and growth parameters. *Sci. Rep.* **2017**, *7*, 11230. [CrossRef]
7. Levin, A.M.; Sitarik, A.R.; Havstad, S.L.; Fujimura, K.E.; Wegienka, G.; Cassidy-Bushrow, A.E.; Kim, H.; Zoratti, E.M.; Lukacs, N.W.; Boushey, H.A.; et al. Joint effects of pregnancy, sociocultural, and environmental factors on early life gut microbiome structure and diversity. *Sci. Rep.* **2016**, *6*, 31775. [CrossRef]
8. Thursby, E.; Juge, N. Introduction to the human gut microbiota. *Biochem. J.* **2017**, *74*, 1823–1836. [CrossRef]
9. Baviera, G.; Leoni, M.C.; Capra, L.; Cipriani, F.; Longo, G.; Maiello, N.; Ricci, G.; Galli, E. Microbiota in healthy skin and in atopic eczema. *BioMed Res. Int.* **2014**, *2014*, 436921. [CrossRef] [PubMed]
10. Capone, K.A.; Dowd, S.E.; Stamatas, G.N.; Nikolovski, J. Diversity of the human skin microbi-ome early in life. *J. Investig. Dermatol.* **2011**, *131*, 2026–2032. [CrossRef]
11. Azad, M.B.; Konya, T.; Guttman, D.S.; Field, C.J.; Sears, M.R.; HayGlass, K.T.; Mandhane, P.J.; Turvey, S.E.; Subbarao, P.; Becker, A.B.; et al. CHILD Study Investigators. Infant gut microbiota and food sensitization: Associations in the first year of life. *Clin. Exp. Allergy* **2015**, *45*, 632–643. [CrossRef] [PubMed]
12. Bunyavanich, S.; Shen, N.; Grishin, A.; Wood, R.; Burks, W.; Dawson, P.; Jones, S.M.; Leung, D.Y.M.; Sampson, H.; Sicherer, S.; et al. Early-life gut microbiome composition and milk allergy resolution. *J. Allergy Clin. Immunol.* **2016**, *138*, 1122–1130. [CrossRef]
13. Łoś-Rycharska, E.; Gołębiewski, M.; Grzybowski, T.; Rogala-Ładniak, U.; Krogulska, A. The microbiome and its impact on food allergy and atopic dermatitis in children. *Adv. Dermatol. Allergol.* **2020**, *5*, 641–650. [CrossRef] [PubMed]
14. Rachid, R.; Stephen-Victor, E.; Chatila, T.A. The microbial origins of food allergy. *J. Allergy Clin. Immunol.* **2021**, *47*, 808–813. [CrossRef] [PubMed]
15. Huang, Y.J.; Marsland, B.J.; Bunyavanich, S.; O'Mahony, L.; Leung, D.Y.M.; Muraro, A.; Fleisher, T.A. The microbiome in allergic disease: Current understanding and future opportunities-2017 PRACTALL document of the American Academy of Allergy, Asthma & Immunology and the European Academy of Allergy and Clinical Immunology. *J. Allergy Clin. Immunol.* **2017**, *139*, 1099–1110. [PubMed]
16. Marrs, T.; Flohr, C. The role of skin and gut microbiota in the development of atopic eczema. *Br. J. Dermatol.* **2016**, *175*, 13–18. [CrossRef] [PubMed]
17. O'Neill, C.A.; Monteleone, G.; McLaughlin, J.T.; Paus, R. The gut-skin axis in health and disease: A paradigm with therapeutic implications. *Bioessays* **2016**, *38*, 1167–1176. [CrossRef]
18. Lee, M.J.; Kang, M.J.; Lee, S.Y.; Lee, E.; Kim, K.; Won, S.; Suh, D.I.; Kim, K.W.; Sheen, Y.H.; Ahn, K.; et al. Perturbations of Gut Microbiome Genes in Infants With Atopic Dermatitis According to Feeding Type. *J. Allergy Clin. Immunol.* **2018**, *141*, 1310–1319. [CrossRef]
19. Koletzko, S.; Niggemann, B.; Arato, A.; Dias, J.A.; Heuschkel, R.; Husby, S.; Mearin, M.L.; Papadopoulou, A.; Ruemmele, F.M.; Staiano, A.; et al. European Society of Pediatric Gastroenterology, Hepatology, and Nutrition. Diagnostic approach and management of cow's milk protein allergy in infants and children: A practical guideline of the GI-committee of ESPGHAN. *J. Pediatr. Gastroenterol. Nutr.* **2012**, *55*, 221–229. [CrossRef] [PubMed]
20. Chen, C.C.; Chen, K.J.; Kong, M.S.; Chang, H.J.; Huang, J.L. Alterations in the gut microbiotas of children with food sensitization in early life. *Pediatr. Allergy Immunol.* **2016**, *27*, 254–262. [CrossRef]
21. Aparicio, M.; Alba, C.; Rodríguez, J.M.; Fernández, L. Microbiological and immunological markers in milk and infant feces for common gastrointestinal disorders: A pilot study. *Nutrients* **2020**, *12*, 634. [CrossRef]
22. Ling, Z.; Li, Z.; Liu, X.; Cheng, Y.; Luo, Y.; Tong, X.; Yuan, L.; Wang, Y.; Sun, J.; Li, L.; et al. Altered fecal microbiota composition associated with food allergy in infants. *Appl. Environ. Microbiol.* **2014**, *80*, 2546–2554. [CrossRef] [PubMed]
23. Song, H.; Yoo, Y.; Hwang, J.; Na, Y.C.; Kim, H.S. Faecalibacterium prausnotzii subspecies-level dysbiosis in the human gut microbiome underlying atopic dermatitis. *J. Allergy Clin. Immunol.* **2016**, *137*, 852–860. [CrossRef] [PubMed]
24. Sicherer, S.H.; Sampson, H.A. Food allergy: A review and update on epidemiology, pathogenesis, diagnosis, prevention, and management. *J. Allergy Clin. Immunol.* **2018**, *141*, 41–58. [CrossRef] [PubMed]
25. Fazlollahi, M.; Chun, Y.; Grishin, A.; Wood, R.A.; Burks, A.W.; Dawson, P.; Jones, S.M.; Leung, D.Y.M.; Sampson, H.A.; Sicherer, S.H.; et al. Early-life gut microbiome and egg allergy. *Allergy* **2018**, *73*, 1515–1524. [CrossRef]
26. Fieten, K.B.; Totté, J.E.E.; Levin, E.; Reyman, M.; Meijer, Y.; Knulst, A.; Schuren, F.; Pasmans, S.G.M.A. Fecal Microbiome and Food Allergy in Pediatric Atopic Dermatitis: A Cross-Sectional Pilot Study. *Int. Arch. Allergy Immunol.* **2018**, *175*, 77–84. [CrossRef]

27. Abrahamsson, T.R.; Jakobsson, H.E.; Andersson, A.F.; Björkstén, B.; Engstrand, L.; Jenmalm, M.C. Low diversity of the gut microbiota in infants with atopic eczema. *J. Allergy Clin. Immunol.* **2012**, *129*, 434–440. [CrossRef]

28. Laborel-Préneron, E.; Bianchi, P.; Boralevi, F.; Lehours, P.; Fraysse, F.; Morice-Picard, F.; Sugai, M. Sato'o, Y.; Badiou, C.; Lina, G.; et al. Effects of the Staphylococcus aureus and Staphylococcus epidermidis Secretomes Isolated from the Skin Microbiota of Atopic Children on CD4+ T Cell Activation. *PLoS ONE* **2015**, *10*, e0141067.

29. Kong, H.H.; Oh, J.; Deming, C.; Conlan, S.; Grice, E.A.; Beatson, M.A.; Nomicos, E.; Polley, C.; Komarow, H.D.; NISC Comparative Se-quence Program; et al. Temporal shifts in the skin microbiome associated with disease flares and treatment in children with atopic dermatitis. *Genome Res.* **2012**, *22*, 850–859. [CrossRef]

30. Stefka, A.T.; Feehley, T.; Tripathi, P.; Qiu, J.; McCoy, K.; Mazmanian, S.K.; Tjota, M.Y.; Seo, G.Y.; Cao, S.; Theriault, B.R.; et al. Commensal bacteria protect against food allergen sensitization. *Proc. Natl. Acad. Sci. USA* **2014**, *111*, 13145. [CrossRef]

31. Noval Rivas, M.; Burton, O.T.; Wise, P.; Charbonnier, L.M.; Georgiev, P.; Oettgen, H.C.; Rachid, R.; Chatila, T.A. Regulatory T cell reprogramming toward a Th2-cell-like lineage impairs oral tolerance and promotes food allergy. *Immunity* **2015**, *42*, 512–523. [CrossRef]

32. Kennedy, E.A.; Connolly, J.; O'B Hourihane, J.; Fallon, P.G.; McLean, W.H.I.; Murray, D.; Jo, J.H.; Segre, J.A.; Kong, H.H.; Irvine, A.D. Skin Microbiome Before Development of Atopic Dermatitis: Early Colonization With Commensal Staphylococci at 2 Months Is Associated With a Lower Risk of Atopic Dermatitis at 1 Year. *J. Allergy Clin. Immunol.* **2017**, *139*, 166–172. [CrossRef]

33. West, C.E.; Rydén, P.; Lundin, D.; Engstrand, L.; Tulic, M.K.; Prescott, S.L. Gut Microbiome and innate immune response patterns in IgE-associated eczema. *Clin. Exp. Allergy* **2015**, *45*, 1419–1429. [CrossRef]

34. Nylund, L.; Nermes, M.; Isolauri, E.; Salminen, S.; de Vos, W.M.; Satokari, R. Severity of atopic disease inversely correlates with intestinal microbiota diversity and butyrate producing bacteria. *Allergy* **2015**, *70*, 241–244. [CrossRef]

35. Li, W.; Xu, X.; Wen, H.; Wang, Z.; Ding, C.; Liu, X.; Gao, Y.; Su, H.; Zhang, J.; Han, Y.; et al. Inverse Association Between the Skin and Oral Microbiota in Atopic Dermatitis. *J. Investig. Dermatol.* **2019**, *139*, 1779–1787. [CrossRef]

36. Asher, M.I.; Keil, U.; Anderson, H.R.; Beasley, R.; Crane, J.; Martinez, F.; Mitchell, E.W.; Pearce, N.; Sibbald, B.; Staward, A.W.; et al. International study of asthma and allergies in childhood (ISAAC): Rationale and methods. *Eur. Respir. J.* **1995**, *8*, 483–491. [CrossRef]

37. Vandenplas, Y.; Dupont, C.; Eigenmann, P.; Host, A.; Kuitunen, M.; Ribes-Koninckx, C.; Shah, N.; Shamir, R.; Staiano, A.; Szajewska, H.; et al. A workshop report on the development of the Cow's Milk-related Symptom Score awareness tool for young children. *Acta Paediatr.* **2015**, *104*, 334–339. [CrossRef]

38. Meyer, R.; Chebar Lozinsky, A.; Fleischer, D.M.; Vieira, M.C.; Du Toit, G.; Vandenplas, Y.; Dupont, C.; Knibb, R.; Uysal, P.; Cavkaytar, O.; et al. Diagnosis and management of Non-IgE gastrointestinal allergies in breastfed infants—An EAACI Position Paper. *Allergy* **2020**, *75*, 14–32. [CrossRef]

39. Schoemaker, A.A.; Sprikkelman, A.B.; Grimshaw, K.E.; Roberts, G.; Grabenhenrich, L.; Rosenfeld, L.; Siegert, S.; Dubakiene, R.; Rudzeviciene, O.; Reche, M.; et al. Incidence and natural history of challenge-proven cow's milk allergy in European children—EuroPrevall birth cohort. *Allergy* **2015**, *70*, 963–972. [CrossRef]

40. Williams, H.C.; Burney, P.G.; Hay, R.J.; Archer, C.B.; Shipley, M.J.; Hunter, J.J.; Bingham, E.A.; Finlay, A.Y.; Pembroke, A.C.; Graham-Brown, R.A.; et al. The U.K. Working Party's Diagnostic Criteria for Atopic Dermatitis. I. Derivation of a minimum set of discriminators for atopic dermatitis. *Br. J. Dermatol.* **1994**, *131*, 383–396. [CrossRef]

41. Williams, H.C. Clinical practice. Atopic dermatitis. *N. Engl. J. Med.* **2005**, *352*, 2314–2324. [CrossRef]

42. Stalder, J.F.; Taïeb, A.; Atherton, D.J.; Bieber, P.; Bonifazi, E.; Broberg, A.; Calza, A.; Coleman, R.; De Prost, Y.; Stalder, J.F.; et al. Severity scoring of atopic dermatitis: The SCORAD index. Consensus Report of the European Task Force on Atopic Dermatitis. *Dermatology* **1993**, *186*, 23–31.

43. Thiem, D.; Gołębiewski, M.; Hulisz, P.; Piernik, A.; Hrynkiewicz, K. How does salinity shape bacterial and fungal microbiomes of Alnus glutinosa Roots? *Front. Microbiol.* **2018**, *9*, 651. [CrossRef] [PubMed]

44. Callahan, B.J.; McMurdie, P.J.; Rosen, M.J.; Han, A.W.; Johnson, A.J.A.; Holmes, S.P. DADA2: High resolution sample inference from Illumina amplicon data. *Nat. Methods* **2016**, *13*, 581–583. [CrossRef] [PubMed]

45. Quast, C.; Pruesse, E.; Yilmaz, P.; Gerken, J.; Schweer, T.; Yarza, P.; Peplies, J.; Glockner, F.O. The SILVA ribosomal RNA gene database project:improved data processing and web-based tools. *Nucleic Acids Res.* **2013**, *41*, D590–D596. [CrossRef]

46. Schloss, P.D.; Westcott, S.L.; Ryabin, T.; Hall, J.R.; Hartmann, M.; Hollister, M.B.; Lesniewski, R.A.; Oakley, B.B.; Parks, D.H.; Robinson, C.J.; et al. Introducing mothur: Open-source, platform-independent, community-supported software for describing and comparing microbial communities. *Appl. Environ. Microbiol.* **2009**, *75*, 7537–7541. [CrossRef]

47. Oksanen, J.; Blanchet, F.G.; Friendly, M.; Kindt, R.; Legendre, P.; McGlinn, D.; Minchin, P.R.; O'Hara, R.B.; Simpson, G.L.; Solymos, P.; et al. Vegan: Community Ecology Package. R Package v.2.5-6. 2019. Available online: https://CRAN.R-project.org/package=vegan (accessed on 23 January 2020).

48. Chen, J. GUniFrac: Generalized UniFrac Distances. R Package v.1.1. 2018. Available online: https://CRAN.R-project.org/package=GUniFrac (accessed on 23 January 2020).

49. Chen, J.; Bittinger, K.; Charlson, E.S.; Hoffmann, C.; Lewis, J.; Wu, G.D.; Collman, R.G.; Bushman, F.D.; Li, H. Associating microbiome composition with environmental covariates using generilized UniFrac distances. *Bioinformatics* **2012**, *28*, 2106–2113. [CrossRef]

50. Sheneman, L.; Evans, J.; Foster, J.A. Clearcut: A fast implementation of relaxed neighbor joining. *Bioinformatics* **2006**, *22*, 2823–2824. [CrossRef]

51. Gołębiewski, M.; Tretyn, A. Generating amplicon reads for microbial community assessment with next-generation sequencing. *J. Appl. Microbiol.* **2020**, *128*, 330–354. [CrossRef]

52. Savage, J.H.; Lee-Sarwar, K.A.; Sordillo, J.; Bunyavanich, S.; Zhou, Y.; O'Connor, G.; Sandel, M.; Bacharier, L.B.; Zeiger, R.; Sodergren, E.; et al. A Prospective Microbiome-Wide Association Study of Food Sensitization and Food Allergy in Early Childhood. *Allergy* **2018**, *73*, 145–152. [CrossRef] [PubMed]

53. Canani, R.B.; De Filippis, F.; Nocerino, R.; Paparo, L.; Di Scala, C.; Cosenza, L.; Della Gatta, G.; Calignano, A.; De Caro, C.; Laiola, M.; et al. Gut Microbiota Composition and Butyrate Production in Children Affected by non-IgE-mediated Cow's Milk Allergy. *Sci. Rep.* **2018**, *8*, 12500. [CrossRef] [PubMed]

54. Park, Y.M.; Lee, S.Y.; Kang, M.J.; Kim, B.S.; Lee, M.J.; Jung, S.S.; Yoon, J.S.; Cho, H.J.; Lee, E.; Yang, S.I.; et al. Imbalance of Gut Streptococcus; Clostridium, and Akkermansia Determines the Natural Course of Atopic Dermatitis in Infant. *Allergy Asthma Immunol. Res.* **2020**, *12*, 322–337. [CrossRef]

55. Kwon, S.; Choi, J.Y.; Shin, J.W.; Huh, C.H.; Park, K.C.; Du, M.H.; Yoon, S.; Na, J.I. Changes in Lesional and Non-lesional Skin Microbiome During Treatment of Atopic Dermatitis. *Acta Derm. Venereol.* **2019**, *99*, 284–290. [CrossRef]

56. Dong, P.; Feng, J.J.; Yan, D.Y.; Lyu, Y.J.; Xu, X. Early-life gut microbiome and cow's milk allergy- a prospective case—Control 6-month follow-up study. *Saudi. J. Biol. Sci.* **2018**, *25*, 875–880. [CrossRef]

57. Breiman, L. Random Forests. *Mach. Learn.* **2001**, *45*, 5–32. [CrossRef]

58. Saito, Y.; Sato, T.; Nomoto, K.; Tsuji, H. Identification of phenol- and p-cresol-producing intestinal bacteria by using media supplemented with tyrosine and its metabolites. *FEMS Microbiol. Ecol.* **2018**, *94*, fiy125. [CrossRef] [PubMed]

59. Hua, X.; Goedert, J.J.; Pu, A.; Yu, G.; Shi, J.S. Allergy associations with the adult fecal microbiota: Analysis of the American Gut Project. *EBioMedicine* **2016**, *3*, 172–179. [CrossRef]

60. Ruohtula, T.; de Goffau, M.C.; Nieminen, J.K.; Honkanen, J.; Siljander, H.; Hämäläinen, A.M.; Peet, A.; Tillmann, V.; Ilonen, J.; Niemelä, O.; et al. Maturation of Gut Microbiota and Circulating Regulatory T Cells and Development of IgE Sensitization in Early Life. *Front. Immunol.* **2019**, *10*, 2494. [CrossRef] [PubMed]

61. Baldwin, H.E.; Bhatia, N.D.; Friedman, A.; Eng, R.M.; Seite, S. The Role of Cutaneous Microbiota Harmony in Maintaining a Functional Skin Barrier. *J. Drugs Dermatol.* **2017**, *16*, 12–18. [CrossRef]

62. Seité, S.; Zelenkova, H.; Martin, R. Clinical efficacy of emollients in atopic dermatitis patients—Relationship with the skin microbiota modification. *Clin. Cosmet. Investig. Dermatol.* **2017**, *10*, 25–33. [CrossRef]

63. Leung, D.Y.M.; Calatroni, A.; Zaramela, L.S.; LeBeau, P.K.; Dyjack, N.; Brar, K.; David, G.; Johnson, K.; Leung, S.; Ramirez-Gama, M.; et al. The nonlesional skin surface distinguishes atopic dermatitis with food allergy as a unique endotype. *Sci. Transl. Med.* **2019**, *11*, eaav2685. [CrossRef] [PubMed]

64. Bosman, E.S.; Albert, A.Y.; Lui, H.; Dutz, J.P.; Vallance, B.A. Skin exposure to narrow band ultraviolet (UVB) light modulates the human intestinal microbiome. *Front. Microbiol.* **2019**, *10*, 2410. [CrossRef] [PubMed]

65. Henning, S.M.; Yang, J.; Lee, R.P.; Huang, J.; Hsu, M.; Thames, G.; Gilbuena, I.; Long, J.; Xu, Y.; Park, E.H.; et al. Pomegranate Juice and Extract Consumption Increases the Resistance to UVB-induced Erythema and Changes the Skin Microbiome in Healthy Women: A Randomized Controlled Trial. *Sci. Rep.* **2019**, *9*, 14528. [CrossRef] [PubMed]

66. Lee, S.Y.; Lee, E.; Park, Y.M.; Hong, S.J. Microbiome in the gut-skin axis in atopic dermatitis. *Allergy Asthma Immunol. Res.* **2018**, *10*, 354–362. [CrossRef]

67. Nam, B.; Kim, S.A.; Park, S.D.; Kim, H.J.; Kim, J.S.; Bae, C.H.; Kim, J.Y.; Nam, W.; Lee, J.L.; Sim, J.H. Regulatory effects of Lactobacillus plantarum HY7714 on skin health by improving intestinal condition. *PLoS ONE* **2020**, *15*, e0231268. [CrossRef]

68. Nagino, T.; Kaga, C.; Kano, M.; Masuoka, N.; Anbe, M.; Moriyama, K.; Maruyama, K.; Nakamura, S.; Shida, K.; Miyazaki, K. Effects of fermented soymilk with Lactobacillus casei Shirota on skin condition and the gut microbiota: A randomised clinical pilot trial. *Benef. Microbes* **2018**, *9*, 209–218. [CrossRef]

69. Zheng, H.; Liang, H.; Wang, Y.; Miao, M.; Shi, T.; Yang, F.; Liu, E.; Yuan, W.; Ji, Z.S.; Li, D.K. Altered Gut Microbiota Composition Associated With Eczema in Infants. *PLoS ONE* **2016**, *11*, e0166026. [CrossRef]

70. Reddel, S.; Del Chierico, F.; Quagliariello, A.; Giancristoforo, S.; Vernocchi, P.; Russo, A.; Fiocchi, A.; Rossi, P.; Putignani, L.; El Hachem, M. Gut Microbiota Profile in Children Affected by Atopic Dermatitis and Evaluation of Intestinal Persistence of a Probiotic Mixture. *Sci. Rep.* **2019**, *9*, 4996. [CrossRef]

71. Zheng, D.; Liwinski, T.; Elinav, E. Interaction between microbiota and immunity in health and disease. *Cell Res.* **2020**, *30*, 492–506. [CrossRef]

72. Clavel, T.; Gomes Neto, J.C.; Lagkouvardos, I.; RamerTait, A.T. Deciphering interactions between the gut microbiota and the immune system via microbial cultivation and minimal microbiomesd. *Immunol. Rev.* **2017**, *279*, 8–22. [CrossRef] [PubMed]

73. Taylor, S.L.; Leong, L.E.X.; Choo, J.M.; Wesselingh, S.; Yang, I.A.; Upham, J.W.; Reynolds, P.N.; Hodge, S.; James, A.L.; Jenkins, C.; et al. Inflammatory phenotypes in patients with severe asthma are associated with distinct airway microbiology. *J. Allergy Clin. Immunol.* **2018**, *41*, 94–103. [CrossRef]

74. Mackenzie, B.W.; Waite, D.W.; Hoggard, M.; Douglas, R.G.; Taylor, M.W.; Biswas, K. Bacterial community collapse: A meta-analysis of the sinonasal microbiota in chronic rhinosinusitis. *Environ. Microbiol.* **2017**, *19*, 381–392. [CrossRef] [PubMed]

75. Chandra, J.; Retuerto, M.; Seité, S.; Martin, R.; Kus, M.; Ghannoum, M.A.; Baron, E.; Mukherjee, P.K. Effect of an Emollient on the Mycobiome of Atopic Dermatitis Patients. *J. Drugs Dermatol.* **2018**, *17*, 1039–1048. [PubMed]
76. Hoyles, L.; Inganas, E.; Falsen, E.; Drancourt, M.; Weiss, N.; McCartney, A.L.; Collins, M.D. Bifidobacterium scardovii sp. nov., from human sources. *Int. J. Syst. Evol. Microbiol.* **2002**, *52*, 995–999. [PubMed]
77. Man, W.H.; Scheltema, N.M.; Clerc, M.; van Houten, M.A.; Nibbelke, E.E.; Achten, N.B.; Arp, K.; Sanders, E.A.M.; Bont, L.J.; Bogaert, D. Infant respiratory syncytial virus prophylaxis and nasopharyngeal microbiota until 6 years of life: A subanalysis of the MAKI randomised controlled trial. *Lancet Respir. Med.* **2020**, *8*, 1022–1031. [CrossRef]
78. Johnson, R.C.; Ellis, M.W.; Lanier, J.B.; Schlett, C.D.; Cui, T.; Merrell, D.S. Correlation between nasal microbiome composition and remote purulent skin and soft tissue infections. *Infect. Immun.* **2015**, *83*, 802–811. [CrossRef] [PubMed]
79. Fernando, J.R.; Butler, C.A.; Adams, G.G.; Mitchell, H.L.; Dashper, S.G.; Escobar, K.; Hoffmann, B.; Shen, P.; Walker, G.D.; Yuan, Y.; et al. The prebiotic effect of CPP-ACP sugar-free chewing gum. *J. Dent.* **2019**, *91*, 103225. [CrossRef]
80. Cukrowska, B.; Bierła, J.B.; Zakrzewska, M.; Klukowski, M.; Maciorkowska, E. The Relationship between the Infant Gut Microbiotaand Allergy. The Role of Bifidobacterium breve and Prebiotic Oligosaccharides in the Activation of Anti-Allergic Mechanisms in Early Life. *Nutrients* **2020**, *12*, 946. [CrossRef]
81. Meisel, J.S.; Hannigan, G.D.; Tyldsley, A.S.; SanMiguel, A.J.; Hodkinson, B.P.; Zheng, Q.; Grice, E.A. Skin Microbiome Surveys Are Strongly Influenced by Experimental Design. *J. Investig. Dermatol.* **2016**, *136*, 947–956. [CrossRef]
82. Stehlikova, Z.; Kostovcik, M.; Kostovcikova, K.; Kverka, M.; Juzlova, K.; Rob, F.; Hercogova, J.; Bohac, P.; Pinto, Y.; Uzan, A.; et al. Dysbiosis of Skin Microbiota in Psoriatic Patients: Co-occurrence of Fungal and Bacterial Communities. *Front. Microbiol.* **2019**, *10*, 438. [CrossRef]

Article

Chronic Milk-Dependent Food Protein-Induced Enterocolitis Syndrome in Children from West Pomerania Region

Karolina Bulsa [1], Małgorzata Standowicz [2], Elżbieta Baryła-Pankiewicz [3] and Grażyna Czaja-Bulsa [4,*]

[1] Szczecin Outpatient Clinic, 71-050 Szczecin, Poland; k.bulsa@gmail.com
[2] Doctoral Studies, Pomeranian Medical University, 70-204 Szczecin, Poland; malgo.kaczorowska@gmail.com
[3] Neonatology Clinic, Pomeranian Medical University, 70-204 Police, Poland; elzpan@gmail.com
[4] Chair and Department of Paediatrics and Paediatric Nursing, Pomeranian Medical University, 70-204 Szczecin, Poland
* Correspondence: grazyna.bulsa@wp.pl; Tel.: +48-91-480-09-51; Fax: +48-91-880-61-46

Abstract: Characteristics of chronic milk-dependent food protein-induced enterocolitis syndrome (FPIES) in children from the region of Western Pomerania were studied. Prospectively, 55 children were diagnosed at a median of 2.2 months. The open food challenges (OFC), morphologies, milk-specific IgE (sIgE) (FEIA method, CAP system), and skin prick tests (SPTs) were examined. Vomiting and diarrhea escalated gradually but quickly led to growth retardation. Of the infants, 49% had BMI < 10 c, 20% BMI < 3 c; 25% had anemia, and 15% had hypoalbuminemia. During the OFCs we observed acute symptoms that appeared after 2–3 h: vomiting diarrhea and pallor. A total of 42% children required intravenous hydration. Casein hydrolysates or amino acids formulae (20%) were used in treatment. In 25% of children, SPT and milk sIgE were found, in 18%—other food SPTs, and in 14% allergy to other foods. A transition to IgE-dependent milk allergy was seen in 3 children. In the twelfth month of life, 62% of children had tolerance to milk, and in the twenty-fifth month—87%. Conclusions. Chronic milk-dependent FPIES resolves in most children. By the age of 2 children are at risk of multiple food sensitization, and those who have milk sIgE are at risk to transition to IgE-mediated milk allergy. Every OFC needs to be supervised due to possible severe reactions.

Keywords: milk allergy; children; non-IgE mediated CMA; food protein-induced enterocolitis syndrome; FPIES

Citation: Bulsa, K.; Standowicz, M.; Baryła-Pankiewicz, E.; Czaja-Bulsa, G. Chronic Milk-Dependent Food Protein-Induced Enterocolitis Syndrome in Children from West Pomerania Region. *Nutrients* **2021**, *13*, 4137. https://doi.org/10.3390/nu13114137

Academic Editor: Carla Mastrorilli

Received: 3 October 2021
Accepted: 17 November 2021
Published: 19 November 2021

Publisher's Note: MDPI stays neutral with regard to jurisdictional claims in published maps and institutional affiliations.

1. Introduction

Cow's milk allergy (CMA) is the most common allergy in the first year of life. It takes two forms: the IgE-mediated CMA (IgE-CMA) and the non-IgE mediated CMA (non-IgE-CMA). Non-IgE-CMA is characterized by digestive symptoms and has a good prognosis, usually resolving before the age of three. In contrast to IgE-CMA, the diagnosis of various non-IgE-CMA syndromes can be challenging due to the overall lack of non-invasive confirmatory testing for these disorders. Many of the non-IgE-CMA syndromes are diagnosed clinically based on history, diagnostic milk-free diet, and followed by positive milk provocation test, which is a "gold standard" for diagnosing these diseases [1,2].

The first classification of non-IgE-CMA gastrointestinal disorders was proposed by Sampson HA in 2003 [3]. The classification was adopted by WAO in 2010 (DRACMA) and EAACI in 2014. [4,5]. It covers three items: food protein-induced enterocolitis syndrome (FPIES), food protein-induced allergic proctocolitis (FPIAP), and food protein-induced enteropathy syndrome (FPIE) as well as the syndrome of eosinophilic gastrointestinal diseases (EGID), where milk can cause allergic reactions under the IgE-dependent and IgE-independent mechanism.

FPIES is a non-IgE cell-mediated food allergy [6]. The first international consensus on these diseases was published in 2017 by an international workgroup convened through the Adverse Reactions to Foods Committee of the AAAAI and the International

FPIES Association advocacy group [7]. Previously, diagnosis had been made based on descriptions [8,9].

FPIES usually manifests with repeated vomiting, and less commonly with watery diarrhea, often accompanied by lethargy and pallor. Severe cases can lead to dehydration with ionic disturbances, acidosis, methemoglobinemia, and hypotension (in at least 15% of reactions) mimicking sepsis. Delayed onset (1–4 h after food ingestion) and absence of cutaneous and respiratory symptoms suggest a systemic reaction different from anaphylaxis [7].

FPIES is a syndrome that occurs in two forms, acute and chronic. The acute form of FPIES is much more severe and is caused by food ingested intermittently or after a period of avoidance (solid foods); therefore, it occurs in infants no sooner than after the introduction of modified diet, i.e., usually after 6 months of life. The foods that most commonly cause acute FPIES are rice and oat, which account for almost $^1/_3$ of cases in the USA and Australia [9–11]. In 2009, Mehr et al. highlighted the emerging importance of rice, a food commonly thought to be "hypoallergenic", which cause severe FPIES [11]. In Spain and Italy, FPIES is often caused by a fish-based diet, which is rare in other countries [7,12,13]. Other foods more likely to cause FPIES symptoms include corn, peas, poultry, egg, and goat milk [7].

The chronic form of FPIES is caused by regularly administered food, typically milk or soy infant formula. It is reported only in infants younger than 4 months of age, usually shortly after the end of breastfeeding (2–3 weeks). The main symptoms of chronic FPIES are intermittent vomiting and watery diarrhea rapidly leading to weight and growth deficits. Severe type of chronic FPIES can lead to dehydration and hypoalbuminemia. During the food oral challenge an acute reaction always occurs in the following order: first vomiting (after 1–4 h of food ingestion) followed by watery diarrhea (after 5–10 h of food ingestion). This acute symptomatology during the oral challenge after food avoidance is typical for chronical FPIES. It is also the basis for distinguishing chronic FPIES from FPIE and eosinophilic gastroenteritis [1,7].

The purpose of the study was to describe chronic milk-dependent FPIES (chronic milk-FPIES) in children from the region of Western Pomerania who were diagnosed over a 5-year period (2014–2018).

2. Materials and Methods

The prospective study was conducted for 7 years (2014–2020). During the first 5 years (2014–2018), chronic milk-FPIES was diagnosed in 57 children. They lived in the region of West Pomerania and were patients of the Paediatric Gastroenterology and Rheumatology Clinic, Gastroenterology Outpatient Clinic or Allergy Outpatient Clinics in Szczecin.

In the last 2 years of study (2019–2020) we continued the observation of the study group and we did not include new patients.

The patients were selected from children with symptoms indicating CMA. The suspicion of CMA was the premise for including a milk-free diet for 2–12 weeks, depending on the symptoms (Figure 1). After the symptoms had resolved or decreased, the milk oral food challenge (OFC) was performed. A negative OFC outcome ruled out CMA. A positive OFC was the basis for CMA recognition. FPIES was diagnosed in these children, with a delayed response during OFC (symptoms occurred above 2 h after milk ingestion), in whom occurred vomiting, pallor, and diarrhea, and often also severe dehydration.

The criteria for including a child in the study were: chronic milk-FPIES, age (up to 4 months), absence of coexisting chronic diseases, and a parental/legal guardian's written consent for the child to participate in the controlled study. Consent also included the storage and publication of the collected data. Only 55 children (33 boys, 60.0%) at the age of 1.6–2.6 months (median 2.2 months) were included in the study. After 19 months of treatment, three patients discontinued their participation in the project (they did not report for control milk provocations). Ultimately, 52 children (94.5%) completed the study.

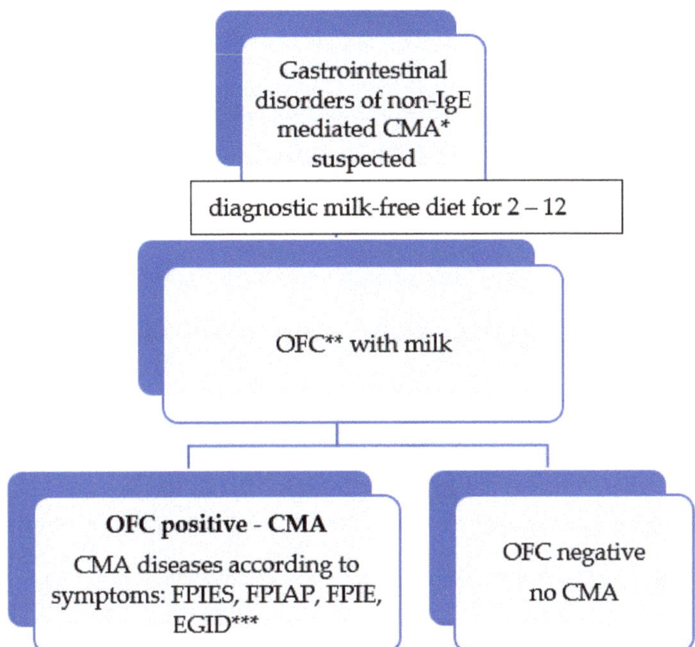

Figure 1. Gastrointestinal disorders of non-IgE mediated CMA diagnosis scheme * CMA—cow's milk allergy; ** OFC—Oral Food Challenge; *** CMA diseases according to symptoms: FPIES—food protein-induced enterocolitis syndrome, FPIAP—food protein-induced proctocolitis, FPIE—food protein-induced enteropathy syndrome, EGID—syndromes of eosinophilic gastrointestinal diseases.

Each patient had a medical and allergic history (recorded recurrent adverse reactions) and underwent physical examination. Every 6 months medical examination was carried out and a follow-up milk OFC was performed. If adverse symptoms occurred after milk ingestion, the provocation was stopped. A positive OFC outcome affirmed the persistence of CMA and was a premise for continuing a milk-free diet. A negative OFC indicated that the child had developed tolerance to milk.

At the time of chronic milk-FPIES diagnosis and during follow-up visits (usually once per year, during the OFC procedure) blood morphology and cow's milk-specific IgE (sIgE) concentration in serum were tested. In addition, skin prick tests (SPTs) with food allergens were performed (Figure 2).

During the diagnostic elimination diet and in the treatment of chronic milk-FPIES, a milk-free diet was administered: either the milk of mothers remaining on a milk-free diet or extensively hydrolyzed infant formulae (eHF; lactose-free casein hydrolysate). Children diagnosed with severe milk allergy received free amino acids formulae (AAF).

The age of chronic milk-FPIES diagnosis was also the age of introduction of a diagnostic milk-free diet. A severe form of CMA was diagnosed according to WAO and ESPGHAN recommendations [4,14]. In all children the diagnosis of FPIES followed the 2017 criteria [7]. Previously, we used the Sicherer et al. criteria [8].

The open OFC procedures were always commenced in hospital conditions, under the control of a nurse and a doctor, with access to anti-shock drugs [15]. After a negative lip test (a drop of milk), gradually higher doses of milk were administered every 15 min (1, 2, 5, 10, 20, 50, and 100 mL). Infants younger than 6 months received at least 100 mL of milk. Patients remained under observation for at least 4–8 h following the end of OFC [15–18]. The provocation was continued at home for the next 6 days. Every day, parents administered the milk mixture corresponding in volume to one meal (older children—up to 250 mL), the information about possible adverse reactions was recorded in the observation card. After 6 days (or earlier, if side effects had occurred), the doctor examined the OFC outcome. In total, the results of 139 milk OFCs were analyzed.

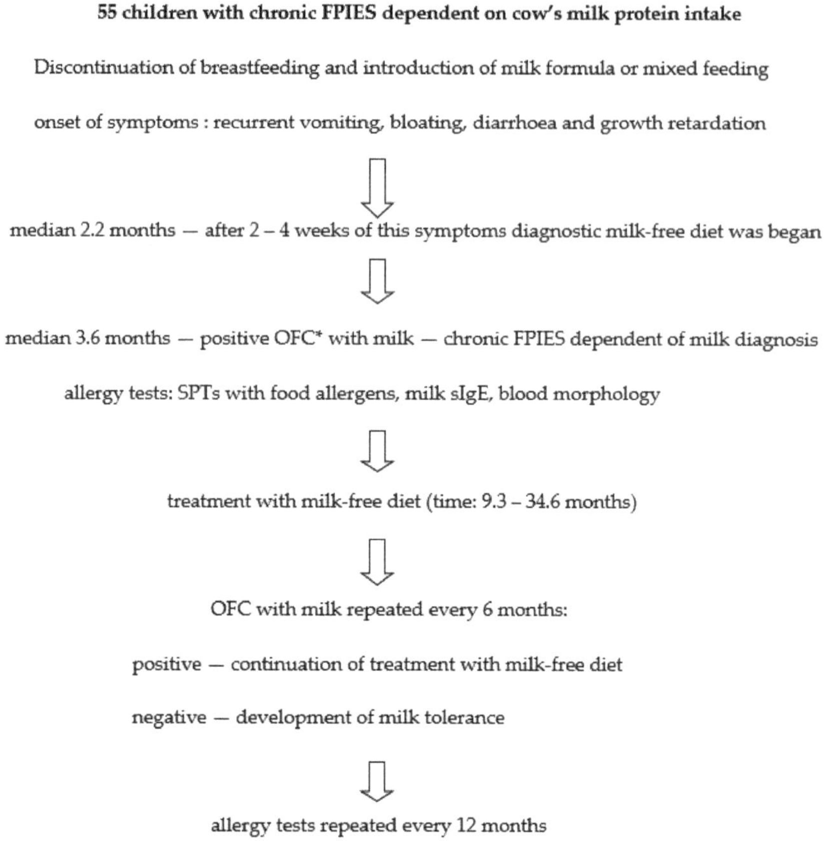

55 children with chronic FPIES dependent on cow's milk protein intake

Discontinuation of breastfeeding and introduction of milk formula or mixed feeding

onset of symptoms : recurrent vomiting, bloating, diarrhoea and growth retardation

median 2.2 months — after 2 – 4 weeks of this symptoms diagnostic milk-free diet was began

median 3.6 months — positive OFC* with milk — chronic FPIES dependent of milk diagnosis

allergy tests: SPTs with food allergens, milk sIgE, blood morphology

treatment with milk-free diet (time: 9.3 – 34.6 months)

OFC with milk repeated every 6 months:

positive — continuation of treatment with milk-free diet

negative — development of milk tolerance

allergy tests repeated every 12 months

Figure 2. Diagnostic and treatment scheme in the study group of children with chronic FPIES dependent on cow's milk protein intake. OFC *—Oral Food Challenge.

During the first OFC all the children had an intravenous entry, while during the next OFC, only those with elevated milk sIgE levels.

Cow's milk sIgE concentration was determined by FEIA method in CAP system with automatic UniCAP apparatus from Phadia. The determination parameters were the range between 0.35–100 kU/L; accuracy between 2–9.1%; sensitivity < 2 kU/L; repeatability at 98%; specificity at 100%. Recommended range: healthy patients < 0.35 kU/L, atopics > 0.35 kU/L.

Skin prick tests (SPTs) with food allergens were performed with a modified method on the skin on the patient's back. The positive control was histamine solution (1 mg/1 mL), while the negative one—a diluent. Commercial Allergopharma and ALK food solutions were used. The wheal and erythema size were measured after 20 min. SPT was regarded positive when the sum of the half of the longest diameter and the midpoint orthogonal diameter of the wheal with allergen was at least equal in diameter to the histamine wheal and at least 3 mm higher than the negative control.

Finally, all data were collected in electronic form in MS Excel spreadsheet and were subject to statistical analysis. Continuous variables were described by median, minimum, and maximum values. Discrete variables were described by their abundance and frequency of occurrence.

The research was approved by the PUM Bioethics Committee No KB-0012/80/14. The research was financed by the statutory activities (WNoZ-319-01/s/12/2013–2020) and by the NCN grant No 2016/21/N/NZ7/03409. The presented results are part of the ongoing project.

3. Results

Chronic FPIES after consumption of cow's milk proteins was diagnosed in 55 children in median age of 2.2 months (1.6–2.6 months) (Table 1). It was the age at which the diagnostic milk-free diet began. Symptoms of CMA appeared after cessation of breastfeeding and introduction of milk formula (80% of children) or mixed feeding (20% of children). After 2–4 weeks of CMA symptoms, a milk-free diet was started. Vomiting and diarrhea escalated gradually but quickly led to physical retardation.

Table 1. Characteristics of children from the study group with chronic food protein-induced enterocolitis syndrome (FPIES) dependent on cow's milk protein intake in the study group of children.

Characteristic	Number of Children ($n = 55$)
Age at onset (median/range)	2.2 months (1.6–2.6 months).
Feeding at the time of onset	Milk formula—80%, breast milk with milk formula—20% (mixed)
Atopic background: -family history of atopy -personal history of atopy	64%: mother—52%, father—58%, siblings—29% 25%
FPIES symptoms: -vomiting -bloating -diarrhea/with mucous/with blood	100% 100% 100%/42%/31%
Body weight: BMI * < 10 c BMI ≤ 3 c	49% 20%
Laboratory findings: -iron deficiency anemia -hypoproteinemia	25% 15%
Allergy tests: -peripheral blood eosinophilia -elevated sIgE and SPT ** for milk -positive SPTs for another foods	Absent 25% 18%
1. OFC *** -time (median/range): Symptoms: -vomiting (2–3 h) -pallor (2–3 h) -diarrhea (4–10 h) -dehydration/intravenous hydration -ondansetron -leukocytosis with neutrophilia	3.6 months (2.1–5.5 months) 100% 100% 100% 42% 7% 80%
Treatment: -lactose-free casein hydrolyzates -amino acids formulae (elementary diet)	80% 20%

* BMI—body mass index; c—percentile; ** SPT—skin prick test; *** OFC—oral food challenge.

At the time of diagnosis as many as half of infants (49%) had low BMI < 10 c (Table 1). In every fifth child it was extremely low at BMI ≤ 3 c. A quarter of the children had iron deficiency anemia and 15% of the children suffered from hypoalbuminemia.

All 55 children with chronic milk-FPIES were affected by the recurrent vomiting, bloating, and diarrhea, sometimes with mucous (42%) and blood in stools (31%) (Table 1). Their weight gain was poor.

The first milk OFC (1. OFC) was performed in children in median age of 3.6 months (2.1–5.5 months). Vomiting (after 2–3 h), diarrhea (after 4–10 h), and pallor were observed in all children (Table 1). As many as 42% of children required intravenous hydration and 7% received ondansetron. Their blood tests revealed increased white blood cell count and neutrophilia in 80% of children. There was no eosinophilia. No child had acidosis or methemoglobinemia. After the diagnosis of chronic milk-FPIES was made, parents were informed that the supply of milk or milk products to their children should only be attempted under medical supervision. The supply of these foods at home can be dangerous.

A family history of atopy was reported in 64% of children (mother—52%, father—58%, or siblings—29%) (Table 1). At the time of diagnosis of chronic milk-FPIES, 18% of the children also had atopic dermatitis (AD).

Lactose-free casein hydrolysates (eHF) and AFF were used in treatment of children with chronic milk-FPIES. AFF was administered to 11 children (20%) who were diagnosed with severe CMA type (BMI $\leq$ 3 c). Resolution of allergy symptoms and improvement in health were observed in all infants already after 3–14 days of dietary treatment.

In every fourth child the elevated sIgE and positive SPT for cow's milk proteins were found (Table 1). Typically, sIgE values were low, in the range of 0.35–0.7 kU/L, rarely higher (2.8, 3.9, 6.9, 28.3 kU/L) and they decreased in the next tests. At the time of chronic milk-FPIES diagnosis no immediate reactions after milk consumption were observed. Immediate symptoms after milk ingestion (IgE-dependent CMA) were seen in 3 children (5.5%) during the subsequent challenge trials (at 19, 25, and 26 months of age) (Table 2). During the provocation, those children developed extensive urticaria, with two of them also suffering from bronchospasm. At that time no symptoms of chronic milk-FPIES were seen in those children. In the following years, their milk sIgE levels increased. Due to lack of parental consent, we could not perform control provocations during 3 years of observation.

Apart from allergy to milk, 10 children (18%) had positive results of SPTs to other foods (egg white, wheat, corn, banana, soybean, peanut), in some cases up to 4 positive SPTs in one child (Table 1). When these foods were first introduced to the diet, two infants developed symptoms of IgE-dependent allergies. One child presented mild urticaria about 30 min after egg white ingestion (at 8 months of age), while second (after 6 months of age)—after wheat intake (Table 3).

3.1. Introducing New Foods into Infants' Diet

According to existing recommendations, parents were suggested to expand the infants' diet after 4 months (2008: EAACI, AAP, 2009—ESPGHAN, 2010—NIAID). Most parents did so (89%), while others began introducing new foods after their babies were 5 months old (Table 4).

All new foods were given at home by parents for 4 days, at a dose consistent with typical intake. The absence of an adverse reaction was the basis for recognizing tolerance to that food and introducing it permanently into the diet [19]. If there were adverse reactions, allergy was diagnosed after contact with a doctor. As the reactions were not severe, the parents repeated the administration of these foods after a few days to check its reproducibility.

The order of products introduced into the diet was changed as recommended by other researchers [8,20]. All parents used a modified dietary expansion method with their children. Pumpkin, broccoli, and cauliflower were served as first foods in a form of a watery mush (Table 4), followed by carrots, potatoes, green beans, zucchini, beets, and parsley. Vegetables with a high risk of causing an allergic reaction, such as sweet potatoes and green peas, were not given until after 6 months of age.

Table 2. Natural history of chronic food protein-induced enterocolitis syndrome (FPIES) dependent on cow's milk protein intake in the study group of children.

Characteristic	Number of Children (*n* = 55)
Age of tolerance to cow's milk development (median/range):	
2. OFC *—12.2 months: 11.5–14.2 months	62%
3. OFC *—19.4 months: 17.4–20.8 months	78% **
4. OFC *—25.2 months: 24.5–26.5 months	87%
5. OFC *—31.1 months: 29.8–31.3 months	87%
6. OFC *—36.8 months: 36.2–37.2 months	89%
Treatment time on a milk-free diet to achieve tolerance (median/range) 16.8 months: 9.3–34.6 months	*n* = 49 children
FPIES transition to IgE-dependent milk allergy:	3 children (5.5%)
3. OFC *	1.8%
4. OFC *	3.6%
Symptoms in children transition to IgE-dependent milk allergy	
-extensive urticaria	5.5%
-bronchospasm	3.6%
Comorbidities:	
-Atopic dermatitis	improvement in half of children
-Hay fever	7%
-Asthma (house dust mite allergy)	5%
Symptoms during the next OFC *:	
-vomiting (2–3 h)	100%
-pallor (2–3 h)	less than 100%
-diarrhea (5–10 h)	less than 100%
-dehydration/only oral hydration	not more than 1/3 of children

* OFC—oral food challenge; ** 3 children dropped out of the study.

After 5 months of age, the diet was expanded by fruits such as peach, grapes, avocado, watermelon, and blueberries. Fruits more frequently causing allergic reactions, i.e., apple, pear, banana, and strawberries were introduced to the diet no sooner than after 6 months of age.

The greatest problems occurred with the supply of cereals, which are given at this age in a form of gruel or porridge. After 6 months of age, the supply of corn and wheat was suggested. Cereals, which in these children most often cause allergy symptoms, i.e., rice and oats, were introduced to the diet between 8 and 10 months of age.

Meat, fish, and egg were given at the same time as is recommended for healthy infants (Table 4). The order of supplying meat types was changed. Rabbit meat was recommended as the first meat, followed by pork and turkey meat. Beef and chicken were not introduced to the diet until the eighth month of age. Soya was not recommended until the twelfth month of age, nor was goat milk because of its high homology to cow's milk.

During the introduction of new foods into the infants' diet, allergy symptoms of IgE-independent allergy were observed (in response to some of the administered foods, despite the delayed time of their administration) (Table 3). Those were skin lesions and/or loose stools occurring from 8 to 24 h after consumption of the harmful food. They occurred in 6 children (11%) in response to 6 foods. The most common foods were: apple (3.6%), rice (3.6%), chicken, and turkey meat (1.8% each).

Table 3. Symptoms of allergy to new foods introduced into the diet in the study group of children with chronic food protein-induced enterocolitis syndrome (FPIES) dependent on cow's milk protein intake.

Characteristic	Number of Children (n = 55)
Foods that triggered IgE-independent allergy symptoms in infants after new food intake *:	6 children (11%)
-apple	2 children (3.6%)
-rice	2 children (3.6%)
-chicken meat	1 child (1.8%)
-turkey meat	1 child (1.8%)
Symptoms of IgE-independent allergy after new food intake *:	
-skin lesions **	4 foods (11%)
-diarrhea	2 foods (3.6%)
Foods that triggered IgE-dependent allergy symptoms (mild urticarial) in infants after new food intake:	2 children (3.6%)
-egg white	1 child (1.8%)
-wheat	1 child (1.8%)
Foods that triggered IgE-dependent allergy symptoms (extensive urticarial) in the next years after new food intake:	3 children (5.5%)
-peanuts	2 children (3.6%)
-soya	1 child (1.8%)

* The provocation was performed twice; ** looked like atopic dermatitis but without itching.

Table 4. Introducing new foods to the diet in the study group of children with chronic food protein-induced enterocolitis syndrome (FPIES) dependent on cow's milk protein intake.

Characteristic	Foods
Age of onset of new foods in the diet:	
-after 4 months	89% of children
-after 5 months	11% of children
Vegetables to be introduced to the diet at 4–6 months of age:	Pumpkin, broccoli, cauliflower then carrots, potatoes, green beans, beets, zucchini Parsley
Fruits to be introduced to the diet after 5 months of age	Peach, grapes, avocado, watermelon, blueberries
Foods to be introduced to the diet after 6 months of age:	
-vegetables	Sweet potatoes and green peas
-fruits	Apple, pear, banana, strawberries
-cereals	Corn, wheat
-meats	Rabbit meat, pork, then turkey meat
-fish	Cod, salmon, and others
Eggs to be introduced to the diet after 7 months of age	First yolk, then egg white
Cereals to be introduced to the diet after 8 months of age	Rice, oats
Meats to be introduced to the diet after 8 months of age	Beef, chicken meat
Foods to be introduced to the diet after 12 months of age	Soya, peanuts

3.2. Course of Chronic Milk-FPIES

The children were observed from 2 to 6 years of age. The weight deficiency present at the time of chronic milk-FPIES diagnosis regressed in all children by the twelfth month of age. Most of them reached a weight and height range of 25–75 c (48 children, 87%). Hypoproteinemia resolved within 2 months, and the treatment of iron deficiency anemia was usually completed before 6 months of age. Blood count abnormalities resolved within 1–2 months.

The chronic milk-FPIES symptoms in the twelfth month of life were absent in 62% of children, in the nineteenth month—in 78% of children (Table 2). The median duration of treatment with a milk-free diet was 16.8 months (range: 9.3–34.6 mo). After 18 months of observation the three of patients discontinued their participation in the project (they did not report for control milk provocations). After 25 months of age, chronic milk-FPIES symptoms persisted only in one child. They disappeared at the thirty-seventh month of age.

During the 3. OFC and 4. OFC three children showed symptoms of IgE-CMA (Table 2). It was the transition of chronic-milk-FPIES to IgE-CMA.

In the next 2–4 years of follow-up, symptoms of IgE-mediated allergy were found following the ingestion of soya (1 child) and peanuts (2 children) (Table 3). All of them developed extensive urticaria.

At the time of diagnosis, 10 children (18%) had mild atopic dermatitis (AD) (Table 2). By the age of 2, it had resolved in 3 children and by the third year in another 2 children (50%). In the following years, 4 children (7%) developed hay fever and 3 children (5%) developed bronchial asthma (house dust mite allergy).

4. Discussion

FPIES was recognized and formally defined in the mid-1970s [21]. Until recently, both acute and chronic FPIES was considered very rare or underdiagnosed [22,23] due to symptoms that can easily be confused with other diseases, especially with sepsis in the instance of acute FPIES [1,7,20]. Intensive research in recent years has shown that FPIES is much more common than it seemed. Studies conducted in centers in Israel and Spain indicated that the cumulative incidence of FPIES in the birth cohort ranges from 0.34% to 0.7% [24,25]. Population studies in the USA have shown that physicians diagnose FPIES in 0.28% of children (<1 and 0.11% of infants) and in 0.22% of adults [26]. Obviously, the data regard two forms of the syndrome and its incidence not only after consumption of milk but of other foods as well.

In the group under study the first symptoms of chronic milk-FPIES occurred shortly after breastfeeding (Table 1). Most mothers decided to continue breastfeeding for a short time, i.e., one month. After 2–4 weeks of infant formula administration, vomiting and diarrhea occurred. The median age of chronic milk-FPIES diagnosis and of the initiation of milk-free diet were low—2.2 months. What is typical for FPIES induced by cow's milk is the appearance of symptoms during formula feeding. Only a few cases of this syndrome have been described in breastfed babies when the allergens were transported with breast milk [27,28]. Researchers in Japan report FPIES in 10% of breastfed babies [29].

The major symptoms of chronic milk-FPIES are intermittent vomiting and watery diarrhea that rapidly lead to weight and growth deficits. In the study group, as many as half of infants had low BMI < 10 c, and in every fifth child the index was extremely low at BMI ≤ 3 c (Table 1).

Although FPIES is a form of IgE-independent food allergy, patients often have atopy, including atopic dermatitis and/or food IgE sensitization. In the study group of children, AD was diagnosed in 18% of children (Table 2). Studies report more frequent AD co-association (31% to 57%) in the United States and Australia while in Korea, Israel, and Italy it is rarer (up to 9%) [7].

In chronic milk-FPIES, food IgE sensitization defined as positive food SPT or serum food sIgE levels occurs in from 4% to 30% of patients [5,7]. In the study group, positive milk SPT and serum milk-sIgE were found in a quarter of children (Table 1). Only in

4 children did the sIgE concentration for milk exceed 0.7 kU/L. Nevertheless, during the 1 OFC in none of them immediate reactions after milk supply were observed. The milk-sIgE concentrations decreased systematically with age. FPIES caused by milk for which there is an elevated sIgE concentration is called atypical FPIES [7,8]. FPIES caused by milk in a child without the presence of milk sIgE is called classical FPIES.

Caubet et al. reported that children with chronic milk-FPIES who had increased milk sIgE were more likely to have persistence of chronic milk-FPIES after 3 years of age compared with those without sIgE. [9]. We did not observe this phenomenon in the study group of children. We also found that the presence of allergies to other foods did not delay the development of milk tolerance.

Only 3 children with elevated milk sIgE levels developed symptoms of IgE-mediated milk allergy during subsequent challenge (Table 2). Thus, a transition of FPIES—IgE-independent milk allergy into IgE-mediated allergy was observed in these children. This is a typical phenomenon in children with FPIES and does not occur in other forms of non-IgE-mediated food allergy. In the study group, we observed such a reaction in 3 children during the third and fourth OFC in 18, 25, and 26 month-olds. During OFC (18–48 mL of milk), all of them developed extensive urticaria. Two of them also developed also bronchospasms (Table 2). In the following years, their sIgE of milk and casein were elevated. Over the 4 years of observation, we did not perform any provocation tests waiting for the period when milk sIgE would start decreasing.

Atopic sensitization to foods other than milk was observed less frequently (25% vs. 18%) (Table 1). Those foods were well tolerated by most children. However, two of them developed symptoms of IgE-mediated allergy (mild urticaria) already in infancy (wheat, egg white) (Table 3). In the following years, extensive urticaria occurred for two more foods—peanuts and soya (Table 3). The coexisting IgE-mediated food allergy at presentation or on follow-up assessment was also reported by other researchers in 20% to 40% of patients [9,11,12]. Children with FPIES are also more likely to be allergic (2–20%) to foods that typically cause FPIES. Cereals (rice, oat), vegetables (sweet potato, green beans, peas), and poultry meats (chicken and turkey), which are typically treated as a potential weak allergen, must be considered in the follow-up of FPIES, particularly in infants with FPIES to cow's milk or soy. Infants with FPIES are at risk of developing hypersensitivity to many food proteins [7,9,24]. In the study group of children, during the introduction of new foods, symptoms of IgE-independent allergy to 6 foods were observed in 6 infants (11% of children). These were rice, apple, turkey, and chicken meat.

Due to significant weight loss, a one-fifth of FPIES patients met the criteria for severe form of CMA and were treated with an elementary diet. The others received lactose-free casein hydrolysates (Table 1). Milk and baked egg were not used during treatment of chronic milk-FPIES primarily because of the short duration of the disease (at median 25 months of age 87% children had already developed tolerance to milk) and the lack of such recommendations in non-atopic allergies [1,7].

For infants with chronic milk-FPIES, avoidance of all forms of milk, including baked and processed milk, is recommended due to a lack of sufficient studies, although tolerance to baked cow's milk has been reported in a small case series [12,20,30]. Some researchers report that some infants with FPIES in OFC tolerate even more than 100 mL of cow's milk [24]. Most researchers believe that the tolerable dose is low [7].

In infants it is recommended to continue breastfeeding. In the study group only 13% of children were breastfed for 6 months, then received eHF.

The introduction of soy milk is not recommended because of the frequent co-occurrence of allergies to both these foods (in about 20–40% of US patients). This coincidence was not observed in patients from other countries such as Italy, Australia, or Israel [12,24,31]. In the study group of children symptoms of allergy to soya occurred only in 1 child in the second year of life in the form of IgE-dependent allergy (urticarial) (Table 3).

Goat and sheep milk are also not recommended in patients with chronic-milk-FPIES due to high homology of the protein sequences in these milks to cow's milk [32].

As regards patients with FPIES, any modification of their diet requires strict supervision, especially when it involves the introduction of grains (rice, oats), poultry and legumes, as they often cause symptoms of acute FPIES, as described in the literature [7,11]. According to the recommendations, in the presented group of children those foods were administered after 8 months of age (Table 4). Simultaneously, meat and fish was included as well. The order of the types of meat to be introduced was changed. Rabbit meat was given first, followed by pork and then turkey and chicken meat and beef. Eggs and pork were given at the same time as it is recommended for children without FPIES (7 and 8 months of age) [1,7].

Despite the delayed time of introducing new foods into the infants' diet, we observed IgE-independent allergy symptoms to 6 foods in 6 children (11%) (Table 3). They were skin lesions or loose stools occurring from 8 to 24 h after consumption of these foods: rice, apple, chicken, and turkey meat. No infant developed FPIES dependent on solid foods.

Although children with chronic milk-FPIES are more likely to react to solid food, most commonly rice or oat, current feeding guidelines do not recommend delay in introducing complementary foods past 6 months of life due to FPIES [5,33,34]. It is recommended that parents introduce a new food as a single ingredient and wait at least 4 days before introducing another food to observe for the development of a reaction [19].

Delayed intake of these foods is more likely to trigger allergy symptoms [5,14]. It is believed that when an infant tolerates the first few foods introduced, dietary expansion may be more tolerable [7]. The use of an elimination diet is always associated with a risk of nutritional deficiencies, so a dietary consultation is recommended [7,35],

OFCs performed in patients with FPIES are high-risk procedures and therefore require medical supervision. In patients with acute FPIES, intensive vomiting quickly leads to dehydration, acidosis, and consciousness disorders that require intensive intravenous treatment. In addition, in patients with chronic FPIES, the supply of harmful food after a period of elimination triggers acute symptoms (e.g., after the first elimination diet period). In all the FPIES patients with elevated milk sIgE levels, who have been treated with a milk-free diet, there is a risk of a sudden reaction during the follow-up OFC.

Therefore, in clinical practice, OFCs are only used in the initial diagnostic evaluation in cases when the patient's history is not clear, symptoms persist despite the elimination of the potential trigger food, or the time course of symptoms is atypical [7]. In an instance of a typical history and improvement on a dairy-free diet, OFC is not performed because the risk of complications following OFC might outweigh its benefits. OFCs are necessary when there is a need to find out whether the child has already developed tolerance to the eliminated food. In chronic FPIES we perform them more often, on average every 6 months, because the disease recedes faster. In acute FPIES, which regresses more slowly, OFCs are performed less frequently, every 12–18 months.

During 1. OFC, vomiting, diarrhea, and pallor occurred in all children in the study group (Table 2). As many as 42% of children required intravenous hydration and 7% received ondansetron. In the next challenge, the percentage of children with diarrhea, pallor, and dehydration decreased. Vomiting was the predominant symptom (Table 1). Dehydration was less frequent and was treated with oral rehydration.

An increase in leukocytosis and neutrophils (>1500 cells/mL) was also observed during OFCs, as reported by other researchers [7,9,21].

During the 3-year observation, milk tolerance developed in 89% of children with chronic milk-FPIES. In the twenty-fifth month of life the proportion was 87%, while at 12 months of life it was recorded at 62% (Table 2). In three children chronic milk-FPIES transformed into IgE-dependent milk allergy. Three other children dropped out of the study after 18 months of follow-up. Similar results were obtained by Hwang at al [36]. Population studies in Israel have shown that 60% of children with cow's milk-induced FPIES develop tolerance by age of 1 year, 75% by age of 2, and 85% by age of 3 [24]. The retrospective US studies revealed that the tolerance was reached by 35% of children by age of 2, 70% by age of 3, and 85% by age of 5 years [10]. The median age of tolerance was 6.7 years. The Korean research shows that tolerance may occur sooner, after 6 or 12 months [36].

5. Conclusions

Chronic-milk-FPIES is a form of IgE-independent allergy. It is a rare syndrome, manifesting itself in the youngest infants after the introduction of milk formula. The symptoms, vomiting and diarrhea, quickly lead to severe growth retardation. Diagnosis is difficult due to the absence of eosinophilia in the blood tests, but the presence of leukocytosis and neutrophilia, which are indicative of infectious diseases. Treatment consists of a milk-free diet. Patients have multiple sensitizations to other foods, both in an IgE-dependent and IgE-independent mechanism. The foods that cause allergy symptoms are different in these two types of allergies. Furthermore, the transition of milk-induced FPIES to IgE-mediated milk allergy may occur in this disease. The occurrence of this transition in the course of OFC poses a risk of severe complications; therefore, milk provocation in these patients always requires intensive medical supervision.

Author Contributions: Conceptualization, G.C.-B. and K.B.; methodology, G.C.-B.; formal analysis, G.C.-B., K.B. and E.B.-P.; data curation, K.B., M.S. and E.B.-P.; writing—original draft preparation, G.C.-B. and K.B.; writing—review and editing, G.C.-B. and E.B.-P.; project administration, G.C.-B. and K.B.; funding acquisition, G.C.-B. and M.S. All authors have read and agreed to the published version of the manuscript.

Funding: National Science Centre, Poland, grant No 2016/21/N/NZ7/03409 and statutory activities of Pomeranian Medical University in Szczecin (WNoZ-319-01/s/12/2013-2020).

Institutional Review Board Statement: Pomeranian Medical University Bioethics Committee No KB-0012/80/14.

Informed Consent Statement: Informed consent was obtained from all subjects involved in the study. Patients consent was waived.

Data Availability Statement: The results of the tests are included in the records of the clinics where the children were treated.

Conflicts of Interest: The authors declare no conflict of interest.

References

1. Nowak-Węgrzyn, A.; Katz, Y.; Mehr, S.S.; Koletzko, S. Non-IgE-mediated gastrointestinal food allergy. *J. Allergy Clin. Immunol.* **2015**, *135*, 1114–1124. [CrossRef] [PubMed]
2. Labrosse, R.; Graham, F.; Caubet, J.C. Non-IgE-mediated gastrointestinal food allergies in children: An update. *Nutrients* **2020**, *12*, 2086. [CrossRef] [PubMed]
3. Sampson, H.A. Food allergy. *J. Allergy Clin. Immunol.* **2003**, *111*, S540–S547. [CrossRef] [PubMed]
4. Fiocchi, A.; Brozek, J.; Schunemann, H.; Bahna, S.; von Berg, A.; Beyer, K.; Bozzola, M.; Bradsher, J.; Compalati, E.; Ebisawa, M.; et al. World Allergy Organization (WAO) diagnosis and rationale for action against cow's milk allergy (DRACMA) guidelines. *World Allergy Organ. J.* **2010**, *3*, 57–161. [CrossRef]
5. Muraro, A.; Werfel, T.; Hoffmann-Sommergruber, K.; Roberts, G.; Beyer, K.; Bindslev-Jensen, C.; Cardona, V.; Dubois, A.; du Toit, G.; Eigenmann, P.; et al. EAACI food allergy and anaphylaxis guidelines: Diagnosis and management of food allergy. *Allergy* **2014**, *69*, 1008–1025. [CrossRef]
6. Boyce, J.A.; Assa'ad, A.; Burks, W.; Jones, S.M.; Sampson, H.A.; Wood, R.A.; Plaut, M.; Cooper, S.F.; Fenton, M.J.; Arshad, S.H.; et al. Guidelines for the diagnosis and management of food allergy in the United States: Summary of the NIAID-sponsored expert panel report. *J. Allergy Clin. Immunol.* **2010**, *126*, 1105–1118. [CrossRef]
7. Nowak-Węgrzyn, A.; Chehade, M.; Groetsch, M.E.; Spergel, J.M.; Wood, R.A.; Allen, K.; Atkins, D.; Bahna, S.; Barad, A.V.; Berin, C.; et al. International consensus guidelines for the diagnosis and management of food protein-induced enterocolitis syndrome: Executive summary—Workgroup report of the adverse reactions to foods committee, American Academy of Allergy, Asthma & Immunology. *J. Allergy Clin. Immunol.* **2017**, *139*, 1111–1126.
8. Sicherer, S.H. Food protein-induced enterocolitis syndrome: Case presentations and management lessons. *J. Allergy Clin. Immunol.* **2005**, *115*, 149–156. [CrossRef]
9. Caubet, J.C.; Ford, L.S.; Sickles, L.; Järvinen, K.M.; Sicherer, S.H.; Sampson, H.A.; Nowak-Węgrzyn, A. Clinical features and resolution of food protein-induced enterocolitis syndrome: 10-year experience. *J. Allergy Clin. Immunol.* **2014**, *134*, 382–389. [CrossRef]

10. Ruffner, M.A.; Ruymann, K.; Barni, S.; Cianferoni, A.; Brown-Whitehorn, T.; Spergel, J.M. Food protein-induced enterocolitis syndrome: Insights from review of a large referral population. *J. Allergy Clin. Immunol. Pract.* **2013**, *1*, 343–349. [CrossRef]
11. Mehr, S.; Kakakios, A.M.; Kemp, A.S. Rice: A common and severe cause of food protein-induced enterocolitis syndrome. *Arch. Dis. Child.* **2009**, *94*, 220–223. [CrossRef] [PubMed]
12. Sopo, S.M.; Giorgio, V.; Iacono, D.I.; Novembre, E.; Mori, F.; Onesimo, R. A multicentre retrospective study of 66 Italian children with food protein-induced enterocolitis syndrome: Different management for different phenotypes. *Clin. Exp. Allergy* **2012**, *42*, 1257–1265. [CrossRef] [PubMed]
13. Vila, L.; García, V.; Rial, M.J.; Novoa, E.; Cacharron, T. Fish is a major trigger of solid food protein-induced enterocolitis syndrome in Spanish children. *J. Allergy Clin. Immunol. Pract.* **2015**, *3*, 621–623. [CrossRef]
14. Koletzko, S.; Niggemann, B.; Arato, A.; Dias, J.A.; Heuschkel, R.; Husby, S.; Mearin, M.L.; Papadopoulou, A.; Ruemmele, F.M.; Staiano, A.; et al. European society of pediatric gastroenterology, hepatology, and nutrition. Diagnostic approach and management of cow-s-milk protein allergy in infants and children: ESPGHAN GI Committee practical guidelines. *J. Pediatr. Gastroenterol. Nutr.* **2012**, *55*, 221–229. [CrossRef]
15. Sampson, H.A.; Gerthvan, W.R.; Bindslev-Jensen, C.; Sicherer, S.; Teuber, S.S.; Burks, A.W. Standardizing double-blind, placebo-controlled oral food challenges: American Academy of Allergy, Asthma & Immunology-European Academy of Allergy and Clinical Immunology PRACTALL consensus report. *J. Allergy Clin. Immunol.* **2012**, *130*, 1260–1274.
16. Nowak-Wegrzyn, A.; Assa'ad, A.H.; Bahna, S.L.; Bock, S.A.; Sicherer, S.H.; Teuber, S.S. Work group report: Oral food challenge testing. *J. Allergy Clin. Immunol.* **2009**, *123*, S365–S383. [CrossRef]
17. Sampson, H.A.; Aceves, S.; Bock, S.A.; James, J.; Jones, S.; Lang, D. Food allergy: A practice parameter update—2014. *J. Allergy Clin. Immunol.* **2014**, *134*, 1016–1025. [CrossRef] [PubMed]
18. Caffarelli, C.; Baldi, F.; Bendandi, B.; Calzone, L.; Marani, M.; Pasquinelli, P. Cow's milk protein allergy in children: A practical guide. *Ital. J. Pediatr.* **2010**, *36*, 5. [CrossRef]
19. Groetch, M.; Henry, M.; Feuling, M.B.; Kim, J. Guidance for the nutrition management of gastrointestinal allergy in pediatrics. *J. Allergy Clin. Immunol. Pract.* **2013**, *1*, 323–331. [CrossRef]
20. Järvinen, K.M.; Nowak-Wegrzyn, A. Food protein-induced enterocolitis syndrome (FPIES): Current management strategies and review of the literature. *J. Allergy Clin. Immunol. Pract.* **2013**, *1*, 317–322. [CrossRef]
21. Powell, G.K. Milk- and soy-induced enterocolitis of infancy: Clinical features and standardization of challenge. *J. Pediatr.* **1978**, *4*, 553–560. [CrossRef]
22. Caubet, J.C.; Szajewska, H.; Shamir, R.; Nowak-Wegrzyn, A. Non-IgE-mediated gastrointestinal food allergies in children. *Pediatr. Allergy Immunol.* **2017**, *28*, 6–17. [CrossRef] [PubMed]
23. Ruffner, M.A.; Spergel, J.M. Non-IgE mediated food allergy syndromes. *Ann. Allergy Asthma Immunol.* **2016**, *117*, 452–454. [CrossRef]
24. Katz, Y.; Goldberg, M.R.; Rajuan, N.; Cohen, A.; Leshno, M. The prevalence and natural course of food protein-induced enterocolitis syndrome to cow's milk: A large-scale, prospective population-based study. *J. Allergy Clin. Immunol.* **2011**, *127*, 647–653. [CrossRef] [PubMed]
25. Alonso, S.B.; Ezquiaga, J.G.; Berzal, P.T.; Tardon, S.D.; San Jose, M.M.; Lopez, P.A.; Bermejo, T.B.; Teruel, S.Q.; Zudaire, L.A.E. Food-protein-induced enterocolitis syndrome: Increased prevalence of this great unknown-results of the PREVALE study. *J. Allergy Clin. Immunol.* **2019**, *143*, 430–433. [CrossRef] [PubMed]
26. Nowak-Wegrzyn, A.; Warren, C.M.; Brown-Whitehorn, T.; Cianferoni, A.; Schultz-Matney, F.; Gupta, R.S. Food protein-induced enterocolitis syndrome in the US population-based study. *J. Allergy Clin. Immunol.* **2019**, *144*, 1128–1130. [CrossRef] [PubMed]
27. Monti, G.; Castagno, E.; Liguori, S.A.; Lupica, M.M.; Tarasco, V.; Viola, S.; Tovo, P.A. Food protein-induced enterocolitis syndrome by cow's milk proteins passed throught breast milk. *J. Allergy Clin. Immunol.* **2011**, *127*, 679–680. [CrossRef] [PubMed]
28. Tan, J.; Campbell, D.; Mehr, S. Food protein-induced enterocolitis syndrome in an exclusively breast-fed infant-an uncommon entity. *J. Allergy Clin. Immunol.* **2012**, *129*, 873–874. [CrossRef]
29. Nomura, I.; Morita, H.; Hosokawa, S.; Hoshina, H.; Fukuie, T.; Watanabe, M.; Ohtsuka, Y.; Shoda, T.; Terada, A.; Takamasu, T.; et al. Four distinct subtypes of non-IgE-mediated gastrointestinal food allergies in neonates and infants, distinguished by their initial symptoms. *J. Allergy Clin. Immunol.* **2011**, *127*, 685–688. [CrossRef] [PubMed]
30. Nowak-Wegrzyn, A.; Berin, M.C.; Mehr, S. Food protein-induced enterocolitis syndrome. *J. Allergy Clin. Immunol. Pract.* **2020**, *8*, 24–35. [CrossRef]
31. NIAID-Sponsored Expert Panel; Boyce, J.A.; Assa'ad, A.; Burks, A.W.; Jones, S.M.; Sampson, H.A.; Wood, R.A.; Plaut, M.; Cooper, S.F.; Fenton, M.J.; et al. Guidelines for the diagnosis and management of food allergy in the United States: Report of the NIAID-sponsored expert panel. *J. Allergy Clin. Immunol.* **2010**, *126*, S1–S58. [CrossRef]
32. Sopo, S.M.; Iacono, I.D.; Greco, M.; Monti, G. Clinical management of food protein-induced enterocolitis syndrome. *Curr. Opin. Allergy Clin. Immunol.* **2014**, *14*, 240–245. [CrossRef]
33. Burks, A.W.; Jones, S.M.; Boyce, J.A.; Sicherer, S.H.; Wood, R.A.; Assa'ad, A.; Sampson, H.A. NIAID-sponsored 2010 guidelines for managing food allergy: Applications in the pediatric population. *Pediatrics* **2011**, *128*, 955–965. [CrossRef] [PubMed]
34. Fleischer, D.M.; Spergel, J.M.; Assa'ad, A.H.; Pongracic, J.A. Primary prevention of allergic disease through nutritional interventions. *J. Allergy Clin. Immunol. Pract.* **2013**, *1*, 29–36. [CrossRef]

35. Venter, C.; Groetch, M. Nutritional management of food protein-induced enterocolitis syndrome. *Curr. Opin. Allergy Clin. Immunol.* **2014**, *14*, 255–262. [CrossRef] [PubMed]
36. Hwang, J.B.; Sohn, S.M.; Kim, A.S. Prospective follow-up oral food challenge in food protein-induced enterocolitis syndrome. *Arch. Dis. Child.* **2009**, *94*, 425–428. [CrossRef] [PubMed]

Article

Food Protein-Induced Enterocolitis Syndrome in Children with Down Syndrome: A Pilot Case-Control Study

Fumiko Okazaki [1], Hiroyuki Wakiguchi [1,*], Yuno Korenaga [1], Kazumasa Takahashi [1], Hiroki Yasudo [1], Ken Fukuda [1], Mototsugu Shimokawa [2] and Shunji Hasegawa [1]

[1] Department of Pediatrics, Yamaguchi University Graduate School of Medicine, 1-1-1 Minamikogushi, Ube, Yamaguchi 755-8505, Japan; fumiko@yamaguchi-u.ac.jp (F.O.); kyuno@yamaguchi-u.ac.jp (Y.K.); p-sama@yamaguchi-u.ac.jp (K.T.); yasudo@yamaguchi-u.ac.jp (H.Y.); kfu10000@yamaguchi-u.ac.jp (K.F.); shunji@yamaguchi-u.ac.jp (S.H.)

[2] Department of Biostatistics, Yamaguchi University Graduate School of Medicine, 1-1-1 Minamikogushi, Ube, Yamaguchi 755-8505, Japan; moto@yamaguchi-u.ac.jp

* Correspondence: hiroyuki@yamaguchi-u.ac.jp; Tel.: +81-836-22-2258; Fax: +81-836-22-2257

Abstract: Food protein-induced enterocolitis syndrome (FPIES) is a non-immunoglobin E-mediated food hypersensitivity disorder. However, little is known about the clinical features of FPIES in patients with Down syndrome (DS). Medical records of children with DS diagnosed at our hospital between 2000 and 2019 were retrospectively reviewed. Among the 43 children with DS, five (11.6%) were diagnosed with FPIES; all cases were severe. In the FPIES group, the median age at onset and tolerance was 84 days and 37.5 months, respectively. Causative foods were cow's milk formula and wheat. The surgical history of colostomy was significantly higher in the FPIES group than in the non-FPIES group. A colostomy was performed in two children in the FPIES group, both of whom had the most severe symptoms of FPIES, including severe dehydration and metabolic acidosis. The surgical history of colostomy and postoperative nutrition of formula milk feeding may have led to the onset of FPIES. Therefore, an amino acid-based formula should be considered for children who undergo gastrointestinal surgeries, especially colostomy in neonates or early infants. When an acute gastrointestinal disease is suspected in children with DS, FPIES should be considered. This may prevent unnecessary tests and invasive treatments.

Keywords: cow's milk allergy; food allergy; food hypersensitivity; gastrointestinal disorder; non-IgE-mediated food hypersensitivity disorder; wheat allergy

Citation: Okazaki, F.; Wakiguchi, H.; Korenaga, Y.; Takahashi, K.; Yasudo, H.; Fukuda, K.; Shimokawa, M.; Hasegawa, S. Food Protein-Induced Enterocolitis Syndrome in Children with Down Syndrome: A Pilot Case-Control Study. *Nutrients* **2022**, *14*, 388. https://doi.org/10.3390/nu14020388

Academic Editor: Carla Mastrorilli

Received: 21 December 2021
Accepted: 13 January 2022
Published: 17 January 2022

Publisher's Note: MDPI stays neutral with regard to jurisdictional claims in published maps and institutional affiliations.

1. Introduction

Food protein-induced enterocolitis syndrome (FPIES) is a non-immunoglobin E (IgE)-mediated food hypersensitivity disorder that primarily affects formula-fed infants and young children [1,2]. The clinical manifestation of FPIES is characterized by profuse and repetitive vomiting, usually occurring within a few hours of feeding, accompanied by lethargy and pallor; diarrhea may also occur within 24 h. Symptoms usually resolve hours after the elimination of the causative food from the diet. Infants who consume foods such as cow's milk or soy-based formula daily may experience chronic weight loss and failure to thrive [3]. FPIES is considered part of a spectrum of allergic diseases that affect only the gut [4]. Although the true incidence of FPIES is not known, large population-based cohort studies from Israel and Spain have reported the cumulative incidence of cow's milk FPIES to be 0.34% and 0.35%, respectively [5,6]. Furthermore, the lifetime prevalence of physician-diagnosed FPIES was reported in the United States, with an estimated prevalence of 0.51% [7]. Because FPIES can be diagnosed clinically and an intestinal biopsy is not performed routinely, little is known about this condition. Specifically, the pathophysiology of FPIES has not been clearly defined and requires further characterization [4].

Several immunological alterations have been reported to be associated with FPIES. Previous studies have suggested the involvement of antigen-specific T cells and their production of proinflammatory cytokines that regulate the permeability of the intestinal barrier [8]. A recent study showed that the levels of transforming growth factor (TGF)-β and interleukin (IL)-10 were significantly lower in children with FPIES than in those with tolerance acquisition [4,6,8]; therefore, TGF-β and IL-10 were proposed as potential biomarkers of FPIES [4,9].

Down syndrome (DS) is caused by a trisomy of human chromosome 21 and occurs in approximately one in 1000 newborns [10–12]. Because the immune system in individuals with DS is altered, accompanied by signs of deficiency and dysregulation, there is a high incidence and prevalence of autoimmune diseases among such individuals [13]. A recent cohort study revealed a lower percentage of allergic sensitization in children with DS than in healthy controls, and no DS children aged 0–2 years had allergic sensitization [14].

We have previously reported two DS children with FPIES [15]. The clinical course of FPIES suggested that this condition may be more severe and require a longer duration to establish tolerance in DS children than in those without DS. However, little is known about the clinical features of FPIES in patients with DS. This study aimed to clarify the clinical features of FPIES in children with DS.

2. Materials and Methods

2.1. Study Population and Data

This was a single-center, retrospective study which was approved by the Institutional Review Board of Yamaguchi University Hospital (H2020-198). Informed consent was obtained from the parents of each patient before their inclusion in the study, and data were collected from patient medical records at our hospital.

Between 1 January 2000 and 31 December 2019, all 62 children with DS born at our hospital or referred to our hospital during the neonatal period were enrolled in this study (Figure 1). Nineteen of the 62 children with DS were excluded from the analysis because they were lost to follow-up. Three children had insufficient data, seven moved shortly after birth, seven were transferred to another hospital for cardiac operations, and two died within the first year of life. Finally, the clinical course of 43 children with DS was followed up for more than 1 year. The DS patients were diagnosed according to chromosomal examinations. FPIES patients were diagnosed according to the criteria of the International Consensus Guidelines for the Diagnosis and Management of FPIES [16].

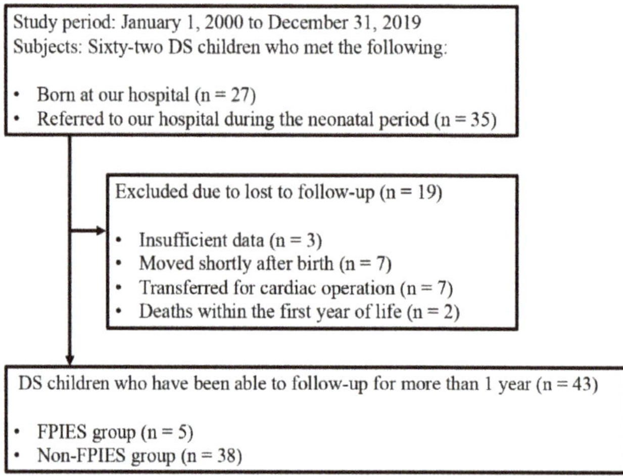

Figure 1. Flow diagram of the subject selection process. DS, Down syndrome; FPIES, food protein-induced enterocolitis syndrome.

2.2. Outcome Measurements

Outcome measurements included demographic characteristics of the FPIES and non-FPIES groups in children with DS and clinical features of FPIES in children with DS. Demographic characteristics included sex, gestational age, birth weight, delivery type, neonatal asphyxia, neonatal jaundice, nutrition, comorbidities, surgical history, serum total IgE level, and antigen-specific IgE level in those below 12 months of age. Serum total IgE level was detected by IgE-LATEX "SEIKEN" (Denka Seiken, Tokyo, Japan), and antigen-specific IgE level was detected by ImmunoCAP (Thermo Fisher Scientific, Uppsala, Sweden). Clinical features of FPIES included age at onset, diagnosis, tolerance, causative foods, clinical symptoms, and severity.

2.3. Statistical Analyses

Demographic characteristics, surgical history, and serum total IgE were summarized using descriptive statistics or contingency tables. The Fisher's exact test or the Mann–Whitney U test was used to compare variables between two groups. Statistical significance was set at $p < 0.05$. Statistical analyses were performed using JMP Pro version 14 (SAS Institute, Cary, NC, USA).

3. Results

Among 43 children with DS, five (11.6%) were diagnosed with FPIES (Figure 1). In the FPIES and non-FPIES groups, sex, gestational age, and median birth weight were approximately the same (Table 1). In the FPIES and non-FPIES groups, more than 60% of children were born by vaginal delivery. Neonatal asphyxia was observed in less than one-fourth, and neonatal jaundice in approximately half of the children. Neonatal asphyxia in this study was defined as an Apgar score ≤ 7 at one min after birth, and neonatal jaundice was defined as requiring phototherapy. There were no breast milk-only infants in either group, and mixed feeding infants accounted for approximately 80% of the total children in both groups.

Table 1. Demographic characteristics of children with DS.

	FPIES (*n* = 5)	Non–FPIES (*n* = 38)	*p*-Value
Male (n (%))	3 (60.0)	20 (52.6)	1.000 †
Median gestational age (weeks (range))	37 (31–38)	38 (30–40)	0.168 ‡
Median birth weight (g (range))	2803 (1714–3224)	2802 (978–3500)	0.910 ‡
Delivery type (n (%))			
Vaginal	3 (60.0)	27 (71.1)	0.630 †
Emergency cesarean section	1 (20.0)	8 (21.1)	1.000 †
Elective cesarean section	1 (20.0)	3 (7.9)	0.402 †
Neonatal asphyxia (n (%))	1/4 (25.0)	5/37 (13.5)	0.483 †
Neonatal jaundice (n (%))	2/4 (50.0)	16/37 (43.2)	1.000 †
Nutrition (n (%))			
Mixed feeding	4 (80.0)	28/32 (87.5)	0.538 †
Formula milk	1 (20.0)	4/32 (12.5)	0.538 †
Breast milk only	0 (0)	0/32 (0)	1.000 †
Comorbidities (n (%))			
Cardiac disease	3 (60.0)	26 (68.4)	1.000 †
ASD	1 (20.0)	7 (18.4)	1.000 †
TOF/DORV	1 (20.0)	5 (13.2)	0.547 †
PDA	0 (0)	4 (10.5)	1.000 †
AVSD	1 (20.0)	3 (7.9)	0.402 †
VSD + PDA	0 (0)	3 (7.9)	1.000 †
VSD	0 (0)	1 (2.6)	1.000 †
VSD + ASD	0 (0)	1 (2.6)	1.000 †
VSD + ASD + PDA	0 (0)	1 (2.6)	1.000 †

Table 1. *Cont.*

	FPIES (*n* = 5)	Non–FPIES (*n* = 38)	*p*-Value
PVP	0 (0)	1 (2.6)	1.000 †
Gastrointestinal disease	2 (40.0)	4 (10.5)	0.136 †
Duodenal atresia	0 (0)	3 (7.9)	1.000 †
Imperforate anus	1 (20.0)	1 (2.6)	0.222 †
Rectovaginal fistula	1 (20.0)	0 (0)	0.116 †
Hematological disorder	1 (20.0)	2 (5.3)	0.316 †
TAM	1 (20.0)	2 (5.3)	0.316 †

ASD, atrial septal defect; AVSD, atrioventricular septal defect, DORV, double outlet right ventricle; DS, Down syndrome; FPIES, food protein-induced enterocolitis syndrome; PDA, patent ductus arteriosus; PVP, pulmonary valve prolapse; TAM, transient abnormal myelopoiesis; TOF, tetralogy of Fallot; VSD, ventricular septal defect. † Fisher's exact test. ‡ Mann–Whitney U test.

Comorbidities of cardiac disease, gastrointestinal disease, and hematological disorder were found in most children with DS in both groups (Table 1). There was no significant difference in comorbidities between the two groups; however, gastrointestinal disease was more common in the FPIES group (40.0% vs. 10.5%, respectively, *p* = 0.136). There was no significant difference in the total surgical history of children with DS between the two groups; however, surgery for gastrointestinal disease was more common in the FPIES group (Table 2). Furthermore, the surgical history of colostomy was significantly higher in the FPIES group than in the non-FPIES group (40.0% vs. 2.6%, respectively, *p* = 0.032).

Table 2. Surgical history of children with DS.

	FPIES (*n* = 5)	Non–FPIES (*n* = 38)	*p*-Value †
Surgical history (n (%))	3 (60.0)	12 (31.6)	0.324
Surgery for cardiac disease (n (%))	1 (20.0)	10 (26.3)	1.000
Intracardiac repair	0 (0)	5 (13.2)	1.000
Ductus arteriosus ligation	0 (0)	3 (7.9)	1.000
BT shunt	0 (0)	2 (5.3)	1.000
PA banding	1 (20.0)	0 (0)	0.116
Surgery for gastrointestinal disease (n (%))	2 (40.0)	4 (10.5) §	0.136
Duodenal atresia repair	0 (0)	3 (7.9) §	1.000
Colostomy	2 (40.0)	1 (2.6)	0.032 *

BT, Blalock-Taussig; DS, Down syndrome; FPIES, food protein-induced enterocolitis syndrome; PA, pulmonary artery. † Fisher's exact test. § Two of them underwent surgery for either intracardiac repair or BT shunt. * Significant at *p* < 0.05.

The serum total IgE levels were determined in all children less than 12 months of age in the FPIES group and six of 38 children in the non-FPIES group (Table 3). The median serum total IgE levels were less than the detection limit (<11 IU/mL) in both groups, and there was no significant difference between the groups. The antigen-specific IgE levels were less than the detection limit (<0.35 kU_A/L) in all children in the FPIES group (Table 4), while they were not determined for the non-FPIES group.

Table 3. Serum total IgE level in children with DS.

	FPIES (*n* = 5)	Non–FPIES (*n* = 6)	*p*-Value
Male (n (%))	3 (60.0)	3 (50.0)	1.000 †
Median age at total IgE test (months (range))	3 (1–12)	10 (1–12)	0.230 ‡
Median total IgE (IU/mL (range))	<11 (<11–11)	<11 (<11–16)	1.000 ‡

DS, Down syndrome; FPIES, food protein-induced enterocolitis syndrome; IgE, immunoglobin E. † Fisher's exact test. ‡ Mann–Whitney U test.

Table 4. Clinical features of FPIES in children with DS.

Participant	1 [15]	2	3	4 [15]	5
Sex	male	male	male	female	female
Age at onset (days)	7	84	321	64	104
Causative foods	CM	CM	wheat	CM	CM
Clinical symptoms					
Vomiting	+	+	+	+	+
Diarrhea	+	+	+	+	+
Bloody stools	−	+	−	+	−
Abdominal distention	+	N/A	N/A	N/A	+
Fever	−	+	+	+	+
Metabolic acidosis	+	+	−	+	+
Dehydration/Shock	−	−	−	+	+
Severity [16]	severe	severe	severe	severe	severe
Comorbidities					
Cardiac disease	−	+ (TOF)	−	+ (ASD)	+ (AVSD)
Gastrointestinal disease	−	−	−	+ (rectovaginal fistula)	+ (imperforate anus)
Hematological disorder	−	−	+ (TAM)	−	−
Surgical history	N/A	PA banding	N/A	colostomy	colostomy
Age at surgery (days)	N/A	50	N/A	63	1
Age at tolerance (months)	132	26	N/A	49	18
Total IgE (IU/mL)	11	<11	<11	<11	<11
Antigen-specific IgE (kU$_A$/L)	<0.35	<0.35	<0.35	<0.35	<0.35

ASD, atrial septal defect; AVSD, atrioventricular septal defect; CM, cow's milk formula; DS, Down syndrome; FPIES, food protein-induced enterocolitis syndrome; IgE, immunoglobin E; N/A, not applicable; PA, pulmonary artery; TAM, transient abnormal myelopoiesis; TOF, tetralogy of Fallot.

In the FPIES group, the median age of onset in five cases was 84 days (Table 4). The causative foods were cow's milk formula in four cases and wheat in one case. Repetitive vomiting and diarrhea were observed in all five cases. Bloody stools and abdominal distension were observed in two cases. Fever and metabolic acidosis were observed in four cases. Severe dehydration was observed in two cases with metabolic acidosis. In the case of severe dehydration, the causative food was cow's milk formula. All five cases were diagnosed as severe according to the guidelines [16], and the patients required infusion.

Cardiac disease was observed in three cases. In Case 2, pulmonary artery banding was performed 50 days after birth for pulmonary hypertension with tetralogy of Fallot. Thirty-four days after the surgery, the brand of formula milk was changed, and vomiting and bloody stool appeared. Intracardiac repair was performed at the age of 12 months for tetralogy of Fallot. In Case 4, the patient did not require surgery for atrial septal defect. In Case 5, intracardiac repair was performed at the age of 19 months for atrioventricular septal defect. Gastrointestinal disease was observed in two cases. In Case 4, colostomy was performed 63 days after birth for rectovaginal fistula. In Case 5, colostomy was performed 1 day after birth for imperforate anus. The postoperative nutrition was formula milk for both cases. In Cases 4 and 5, the closure of colostomy was performed at the age of 23 months and 16 months, and tolerance was acquired at 49 months and 18 months of age, respectively. In Case 3, there were multiple episodes of repetitive vomiting after ingestion of wheat food, such as "udon" noodles and pancakes, at 10 months of age. Case 3 was a recent case in which the parent did not agree to the second oral food challenge; therefore, we could not confirm the acquisition of tolerance. In Case 1, the oral food challenge was performed six times, and finally, it took 11 years for the subject to acquire tolerance to cow's milk formula. The median age of tolerance in the four cases was 37.5 months.

4. Discussions

This report describes the clinical features of FPIES in children with DS. In this study, no significant differences were seen in the total surgical history between the two groups; however, surgery for gastrointestinal disease was more common in the FPIES group. Furthermore, the surgical history of colostomy was significantly higher in the FPIES group than in the non-FPIES group (Table 2). In Cases 4 and 5 (Table 4), surgery for colostomy and postoperative nutrition of formula milk feeding may have led to the onset of FPIES. In Case 4, shortly after surgery, the nutrition of formula milk caused serious symptoms, including severe dehydration and metabolic acidosis. However, in Case 5, formula feeding was resumed after surgery, and repetitive vomiting was observed after 4 months. Subsequently, 7 months after the surgery, watery diarrhea appeared following the administration of antibacterial agents after cardiac catheterization, which caused shock. Both cases were the most severe in this study, and aggressive intervention was required. Finally, the stool form was normalized by the administration of an amino-based formula. In neonates and infants, formula milk after surgery was a risk factor of non-IgE-mediated gastrointestinal food allergies when compared to breast milk [17]. Therefore, an amino acid-based formula should be considered for children who undergo gastrointestinal surgeries, especially colostomy in neonates or early infants. In our study, there were no breast milk-only infants in either group, and mixed feeding infants accounted for approximately 80% of the total children in both groups. The mothers had been instructed that they could breastfeed or formula feed, but a retrospective review of the medical records showed that no children in either group were fully breastfed. We believe that a prospective study is needed to determine whether the active recommendation of full breastfeeding could reduce the incidence of FPIES. In addition, the median serum total IgE was less than the detection limit (<11 IU/mL) in both groups (Table 3). The serum IgE sensitization in infants with DS was low, which was the same as previously reported [14]. This suggests that non-IgE-mediated food hypersensitivity disorder is more likely to occur in DS during infancy.

In this study, all the cases were severe (Table 4). Repetitive vomiting and diarrhea were observed in all cases. Metabolic acidosis and severe dehydration were observed in four patients whose causative food was cow's milk formula. This suggests that cow's milk allergy may cause more serious symptoms than wheat allergy in FPIES patients with DS. There have been no reports of wheat-induced FPIES in patients with DS, besides Case 3 (Table 4). In Case 1 (Table 4), the patient took more than 10 years to acquire tolerance to cow's milk. Immune disorders in DS may be associated with high incidence, severity, and difficulty in the acquisition of tolerance.

The CD4+ CD25+ Foxp3+ regulatory T (Treg) cells, which account for almost 10% of peripheral CD4+ T cells, are essential for the balance between pro- and anti-inflammatory responses at mucosal surfaces. There are two subsets of Treg cells, natural Treg (nTreg) cells and induced Treg (iTreg) cells [18]. While nTreg cells are generated in the thymus, iTreg cells arise from peripheral naïve T cells [18]. The thymus in patients with DS presents profound anatomical and architectural abnormalities [19], which may cause alterations in the maturation process of nTreg cells [20]. Individuals with DS manifest with an over-expressed peripheral nTreg population with a defective inhibitory activity that may partially explain the increased frequency of autoimmune diseases [21]. In a recent study, a higher proportion of circulating nTreg cells specific for cow's milk protein was revealed in infants who had outgrown cow's milk FPIES, suggesting that mucosal induction of tolerance against dietary antigens was associated with the development of nTreg cells [20]. Under steady-state conditions, TGF-β and IL-10 maintain peripheral and gut tolerance [22]. Children with active FPIES against cow's milk have deficient T cell-mediated TGF-β response to casein; therefore, TGF-β could be a promising biomarker for identifying children who are likely to experience FPIES reactions to this allergen [9]. These results suggest that the suppressive action of cow's milk-specific nTreg contributes to the production of TGF-β in children with resolved FPIES to cow's milk [8]. TGF-β is a pleiotropic cytokine that is best known for its regulatory activity and induction of peripheral tolerance [22]. Unlike most other cytokines,

TGF-β is produced by many immune and non-immune cells, and virtually, all cell types are responsive to this pleiotropic cytokine [23]. Similarly, IL-10 levels are significantly lower in patients with cow's milk FPIES; however, the levels tend to increase in children with resolved FPIES to cow's milk [4]. IL-10 is a key regulator of the immune system that acts by limiting the inflammatory response, which could otherwise cause tissue damage and is essential for the homeostasis of the immune system, especially in the gastrointestinal tract [22]. Therefore, increased IL-10 expression is also associated with tolerance acquisition in patients with FPIES [24]. IL-10 is produced mainly by T helper 2 cells, T helper 1 cells, nTreg cells, and natural killer T cells during chronic antigen stimulation [25]. In the absence of effector cytokines and in the presence of high concentrations of TGF-β, naïve CD4+ T cells are converted into iTreg cells that produce TGF-β and IL-10. Considering this information, individuals with DS, who have deficient nTreg inhibitory activity and reduced inhibitory activity of effector cytokines, may be more likely to develop FPIES and with more severity; thus, patients may take a longer time to acquire tolerance. In addition, humoral responses have been investigated in FPIES. Lower levels of antigen-specific IgE, IgA, and IgG4 have been found in patients with FPIES compared with those in patients with resolved FPIES [4]. Therefore, humoral immune responses may also be involved in the pathophysiology of FPIES.

FPIES in adults has been reported for many years; however, only recently, adult case series have been published in the peer-reviewed literature [26–30]. The dramatic symptoms of acute FPIES are usually triggered by shellfish and fish, whereas more chronic gastrointestinal symptoms have been attributed to cow's milk, wheat/gluten, and eggs. The predominance of female adult patients with FPIES (88%) is striking [31] and has been also reported in some reports [27–29]. Contrastingly, infantile FPIES is slightly more common in males [31]. Also, there are no reports of FPIES in adult patients with DS. Further large multicenter studies are needed to better characterize adult FPIES.

The limitations of our study include single-center experience, small sample size, and limited follow-up period. Furthermore, due to the retrospective nature of the study, mild cases of this condition may have been overlooked. In addition, detailed cytokine profiles were not sufficiently examined in this study. Larger prospective multinational cohort studies are required to better understand the true incidence, risk factors, and clinical features of FPIES in patients with DS.

5. Conclusions

In our study, five (11.6%) of the 43 children with DS were diagnosed with FPIES, and all cases were severe. The surgical history of colostomy and postoperative nutrition of formula milk feeding may have led to the onset of FPIES; furthermore, cases involving colostomy were the most severe ones in our study. Therefore, an amino acid-based formula should be considered for children who undergo gastrointestinal surgeries, especially colostomy in neonates or early infants. Serum IgE sensitization in infants with DS was low, as previously reported; thus, non-IgE-mediated food hypersensitivity disorder is more likely to occur in DS during infancy. When an acute gastrointestinal disease is suspected in children with DS, FPIES should be considered. This may prevent performing unnecessary tests and invasive treatments.

Author Contributions: Conceptualization, H.W., F.O., H.Y. and S.H.; methodology, F.O. and H.W.; software, F.O.; validation, H.W. and F.O.; formal analysis, F.O., H.W. and M.S.; investigation, F.O., H.W., Y.K, K.F. and K.T.; resources, F.O., H.W., Y.K., K.F., K.T. and M.S.; data curation, F.O., H.W., Y.K., K.T., K.F. and M.S.; writing—original draft preparation, F.O. and H.W.; writing—review and editing, S.H., F.O., H.W. and H.Y.; visualization, F.O. and H.W.; supervision, S.H., H.W. and H.Y.; project administration, H.W. and F.O. All authors have read and agreed to the published version of the manuscript.

Funding: This work was supported by JSPS KAKENHI (Grant number, JP18K07848) (K.F.), a grant from the Kawano Masanori Memorial Public Interest Incorporated Foundation for Promotion of Pediatrics (Grant number, 27-13) (S.H.), the Morinaga Foundation for Health & Nutrition (K.F.), and Grant-in-Aid for Translational Research of Yamaguchi University Hospital 2018 (S.H.).

Institutional Review Board Statement: The study was conducted in accordance with the Declaration of Helsinki. This study was approved by the Institutional Review Board of Yamaguchi University Hospital (H2020-198).

Informed Consent Statement: Informed consent was obtained from the parents of the patients.

Data Availability Statement: The data presented in this study are available on request from the corresponding author. The data are not publicly available due to privacy.

Acknowledgments: We thank the patients, their families, and their friends for their cooperation.

Conflicts of Interest: The authors declare no conflict of interest.

References

1. Nowak-Węgrzyn, A.; Katz, Y.; Mehr, S.S.; Koletzko, S. Non-IgE-mediated gastrointestinal food allergy. *J. Allergy Clin. Immunol.* **2015**, *135*, 1114–1124. [CrossRef]
2. Powell, G.K. Food protein-induced enterocolitis of infancy: Differential diagnosis and management. *Compr. Ther.* **1986**, *12*, 28–37.
3. Järvinen, K.M.; Nowak-Węgrzyn, A. Food protein-induced enterocolitis syndrome (FPIES): Current management strategies and review of the literature. *J. Allergy Clin. Immunol. Pract.* **2013**, *1*, 317–322. [CrossRef]
4. Caubet, J.C.; Bencharitiwong, R.; Ross, A.; Sampson, H.A.; Berin, M.C.; Nowak-Węgrzyn, A. Humoral and cellular responses to casein in patients with food protein-induced enterocolitis to cow's milk. *J. Allergy Clin. Immunol.* **2017**, *139*, 572–583. [CrossRef]
5. Katz, Y.; Goldberg, M.R.; Rajuan, N.; Cohen, A.; Leshno, M. The prevalence and natural course of food protein-induced enterocolitis syndrome to cow's milk: A large-scale, prospective population-based study. *J. Allergy Clin. Immunol.* **2011**, *127*, 647–653. [CrossRef]
6. Alonso, S.B.; Ezquiaga, J.G.; Berzal, P.T.; Tardón, S.D.; San José, M.M.; López, P.A.; Bermejo, T.B.; Teruel, S.Q.; Echeverría Zudaire, L.Á. Food protein-induced enterocolitis syndrome: Increased prevalence of this great unknown-results of the PREVALE study. *J. Allergy Clin. Immunol.* **2019**, *143*, 430–433. [CrossRef] [PubMed]
7. Nowak-Wegrzyn, A.; Warren, C.M.; Brown-Whitehorn, T.; Cianferoni, A.; Schultz-Matney, F.; Gupta, R.S. Food protein-induced enterocolitis syndrome in the US population-based study. *J. Allergy Clin. Immunol.* **2019**, *144*, 1128–1130. [CrossRef]
8. Caubet, J.C.; Nowak-Węgrzyn, A. Current understanding of the immune mechanisms of food protein-induced enterocolitis syndrome. *Expert Rev. Clin. Immunol.* **2011**, *7*, 317–327. [CrossRef]
9. Konstantinou, G.N.; Bencharitiwong, R.; Grishin, A.; Caubet, J.C.; Bardina, L.; Sicherer, S.H.; Sampson, H.A.; Nowak-Węgrzyn, A. The role of casein-specific IgA and TGF-β in children with food protein-induced enterocolitis syndrome to milk. *Pediatr. Allergy Immunol.* **2014**, *25*, 651–656. [CrossRef] [PubMed]
10. Hoshi, N.; Hattori, R.; Hanatani, K.; Okuyama, K.; Yamada, H.; Kishida, T.; Yamada, T.; Sagawa, T.; Sumiyoshi, Y.; Fujimoto, S. Recent trends in the prevalence of Down syndrome in Japan, 1980–1997. *Am. J. Med. Genet.* **1999**, *84*, 340–345. [CrossRef]
11. Shin, M.; Besser, L.M.; Kucik, J.E.; Lu, C.; Siffel, C.; Correa, A.; Congenital anomaly multistate prevalence and survival collaborative. Prevalence of Down syndrome among children and adolescents in 10 regions of the United States. *Pediatrics* **2009**, *124*, 1565–1571. [CrossRef]
12. Loane, M.; Morris, J.K.; Addor, M.C.; Arriola, L.; Budd, J.; Doray, B.; Garne, E.; Gatt, M.; Haeusler, M.; Khoshnood, B.; et al. Twenty-year trends in the prevalence of Down syndrome and other trisomies in Europe: Impact of maternal age and prenatal screening. *Eur. J. Hum. Genet.* **2013**, *21*, 27–33. [CrossRef] [PubMed]
13. Bull, M.J.; Committee on Genetics. Health supervision for children with Down syndrome. *Pediatrics* **2011**, *128*, 393–406. [CrossRef] [PubMed]
14. Eijsvoogel, N.B.; Hollegien, M.I.; Bok, L.A.; Derksen-Lubsen, G.; Dikken, F.P.J.; Leenders, A.C.A.P.; Pijning, A.; Post, E.D.M.; Wojciechowski, M.; Schmitz, R.; et al. Lower percentage of allergic sensitization in children with Down syndrome. *Pediatr. Allergy Immunol.* **2017**, *28*, 852–857. [CrossRef] [PubMed]
15. Wakiguchi, H.; Hasegawa, S.; Kaneyasu, H.; Kajimoto, M.; Fujimoto, Y.; Hirano, R.; Katsura, S.; Matsumoto, K.; Ichiyama, T.; Ohga, S. Long-lasting non-IgE-mediated gastrointestinal cow's milk allergy in infants with Down syndrome. *Pediatr. Allergy Immunol.* **2015**, *26*, 821–823. [CrossRef]
16. Nowak-Węgrzyn, A.; Chehade, M.; Groetch, M.E.; Spergel, J.M.; Wood, R.A.; Allen, K.; Atkins, D.; Bahna, S.; Barad, A.V.; Berin, C.; et al. International consensus guidelines for the diagnosis and management of food protein-induced enterocolitis syndrome: Executive summary-Workgroup Report of the Adverse Reactions to Foods Committee, American Academy of Allergy, Asthma & Immunology. *J. Allergy Clin. Immunol.* **2017**, *139*, 1111–1126.
17. Korai, T.; Kouchi, K.; Takenouchi, A.; Matsuoka, A.; Yabe, K.; Nakata, C. Neonates undergoing gastrointestinal surgery have a higher incidence of non-IgE-mediated gastrointestinal food allergies. *Pediatr. Surg. Int.* **2018**, *34*, 1009–1017. [CrossRef]

18. Schmitt, E.G.; Williams, C.B. Generation and function of induced regulatory T cells. *Front. Immunol.* **2013**, *4*, 152. [CrossRef]
19. Levin, S.; Schlesinger, M.; Handzel, Z.; Hahn, T.; Altman, Y.; Czernobilsky, B.; Boss, J. Thymic deficiency in Down's syndrome. *Pediatrics* **1979**, *63*, 80–87. [CrossRef]
20. Karlsson, M.R.; Rugtveit, J.; Brandtzaeg, P. Allergen-responsive CD4+CD25+ regulatory T cells in children who have outgrown cow's milk allergy. *J. Exp. Med.* **2004**, *199*, 1679–1688. [CrossRef] [PubMed]
21. Pellegrini, F.P.; Marinoni, M.; Frangione, V.; Tedeschi, A.; Gandini, V.; Ciglia, F.; Mortara, L.; Accolla, R.S.; Nespoli, L. Down syndrome, autoimmunity and T regulatory cells. *Clin. Exp. Immunol.* **2012**, *169*, 238–243. [CrossRef]
22. Sanjabi, S.; Zenewicz, L.A.; Kamanaka, M.; Flavell, R.A. Anti-inflammatory and pro-inflammatory roles of TGF-beta, IL-10, and IL-22 in immunity and autoimmunity. *Curr. Opin. Pharmacol.* **2009**, *9*, 447–453. [CrossRef]
23. Li, M.O.; Wan, Y.Y.; Sanjabi, S.; Robertson, A.K.; Flavell, R.A. Transforming growth factor-beta regulation of immune responses. *Ann. Rev. Immunol.* **2006**, *24*, 99–146. [CrossRef] [PubMed]
24. Mori, F.; Barni, S.; Cianferoni, A.; Pucci, N.; de Martino, M.; Novembre, E. Cytokine expression in CD3+ cells in an infant with food protein-induced enterocolitis syndrome (FPIES): Case report. *Clin. Dev. Immunol.* **2009**, *2009*, 679381. [CrossRef] [PubMed]
25. Motomura, Y.; Kitamura, H.; Hijikata, A.; Matsunaga, Y.; Matsumoto, K.; Inoue, H.; Atarashi, K.; Hori, S.; Watarai, H.; Zhu, J.; et al. The transcription factor E4BP4 regulates the production of IL-10 and IL-13 in CD4+ T cells. *Nat. Immunol* **2011**, *12*, 450–459. [CrossRef] [PubMed]
26. Fernandes, B.N.; Boyle, R.J.; Gore, C.; Simpson, A.; Custovic, A. Food protein-induced enterocolitis syndrome can occur in adults. *J. Allergy Clin. Immunol.* **2012**, *130*, 1199–1200. [CrossRef] [PubMed]
27. Gleich, G.J.; Sebastian, K.; Firszt, R.; Wagner, L.A. Shrimp allergy: Gastrointestinal symptoms commonly occur in the absence of IgE sensitization. *J. Allergy Clin. Immunol. Pract.* **2016**, *4*, 316–318. [CrossRef]
28. Du, Y.J.; Nowak-Wegrzyn, A.; Vadas, P. FPIES in adults. *Ann. Allergy Asthma Immunol.* **2018**, *121*, 736–738. [CrossRef]
29. Tan, J.A.; Smith, W.B. Non-IgE-mediated gastrointestinal food hypersensitivity syndrome in adults. *J. Allergy Clin. Immunol. Pract.* **2014**, *2*, 355–357. [CrossRef]
30. Gonzalez-Delgado, P.; Caparrós, E.; Moreno, M.V.; Cueva, B.; Fernández, J. Food protein-induced enterocolitis-like syndrome in a population of adolescents and adults caused by seafood. *J. Allergy Clin. Immunol. Pract.* **2019**, *7*, 670–672. [CrossRef] [PubMed]
31. Caubet, J.C.; Cianferoni, A.; Groetch, M.; Nowak-Wegrzyn, A. Food protein-induced enterocolitis syndrome. *Clin. Exp. Allergy* **2019**, *49*, 1178–1190. [CrossRef] [PubMed]

Review

Heiner Syndrome and Milk Hypersensitivity: An Updated Overview on the Current Evidence

Stefania Arasi [1,*,†,‡], Carla Mastrorilli [2,†,‡], Luca Pecoraro [3,4,‡], Mattia Giovannini [5,‡], Francesca Mori [5,‡], Simona Barni [5,‡], Lucia Caminiti [6,‡], Riccardo Castagnoli [7,‡], Lucia Liotti [8,‡], Francesca Saretta [9,‡], Gian Luigi Marseglia [7,‡] and Elio Novembre [10,‡]

1. Translational Research in Pediatric Specialities Area, Division of Allergy, Bambino Gesù Children's Hospital, IRCCS, Piazza Sant'Onofrio, 4, 00165 Rome, Italy
2. Pediatric Unit and Emergency, University Hospital Consortium Corporation Polyclinic of Bari, Pediatric Hospital Giovanni XXIII, 70126 Bari, Italy; carla.mastrorilli@icloud.com
3. Department of Medicine, University of Verona, 37129 Verona, Italy; luca.pecoraro@assst-mantova.it
4. Pediatric Unit, ASST Mantua, 46100 Mantua, Italy
5. Allergy Unit, Department of Pediatrics, Meyer Children's University Hospital, 50139 Florence, Italy; mattia.giovannini@unifi.it (M.G.); francesca.mori@meyer.it (F.M.); simona.barni@meyer.it (S.B.)
6. Department of Human Pathology in Adult and Development Age "Gaetano Barresi", Allergy Unit, Department of Pediatrics, AOU Policlinico Gaetano Martino, 98158 Messina, Italy; lucycaminiti@yahoo.it
7. Department of Pediatrics, Pediatric Clinic, Fondazione IRCCS Policlinico San Matteo, University of Pavia, 27100 Pavia, Italy; riccardo.castagnoli@yahoo.it (R.C.); gl.marseglia@smatteo.pv.it (G.L.M.)
8. Department of Pediatrics, AOU Ospedali Riuniti Ancona, Presidio Ospedaliero di Alta Specializzazione "G. Salesi", 60126 Ancona, Italy; lucia.liotti@ospedaliriuniti.marche.it
9. Pediatric Department, Latisana-Palmanova Hospital, Azienda Sanitaria Universitaria Friuli Centrale, 33100 Udine, Italy; francescasaretta@gmail.com
10. Department of Health Science, University of Florence, 70126 Florence, Italy; elio.novembre@meyer.it
* Correspondence: stefania.arasi@opbg.net; Tel.: +39-06-68593570
† These authors contributed equally to this work.
‡ From the Rare Allergic Diseases Commission of the Italian Society of Pediatric Allergy and Immunology.

Citation: Arasi, S.; Mastrorilli, C.; Pecoraro, L.; Giovannini, M.; Mori, F.; Barni, S.; Caminiti, L.; Castagnoli, R.; Liotti, L.; Saretta, F.; et al. Heiner Syndrome and Milk Hypersensitivity: An Updated Overview on the Current Evidence. *Nutrients* **2021**, *13*, 1710. https://doi.org/10.3390/nu13051710

Academic Editor: Yvan Vandenplas

Received: 8 April 2021
Accepted: 17 May 2021
Published: 18 May 2021

Publisher's Note: MDPI stays neutral with regard to jurisdictional claims in published maps and institutional affiliations.

Abstract: Infants affected by Heiner syndrome (HS) display chronic upper or lower respiratory tract infections, including otitis media or pneumonia. Clinically, gastrointestinal signs and symptoms, anemia, recurrent fever and failure to thrive can be also present. Chest X-rays can show patchy infiltrates miming pneumonia. Clinical manifestations usually disappear after a milk-free diet. The pathogenetic mechanism underlying HS remains unexplained, but the formation of immune complexes and the cell-mediated reaction have been proposed. Patients usually outgrow this hypersensitivity within a few years. The aim of this review is to provide an updated overview on the current evidence on HS in children, with a critical approach on the still undefined points of this interesting disease. Finally, we propose the first structured diagnostic approach for HS.

Keywords: allergy; anemia; cow's milk; children; immunology; non-IgE-mediated food allergy; pneumonia; pulmonary hemosiderosis; pulmonary infiltrates

1. Introduction

Heiner syndrome (HS) is a rare food-induced hypersensitivity disease characterized by chronic respiratory symptoms with X-ray (XR) infiltrates, and the resolution of signs and symptoms after the removal of milk proteins. Other clinical manifestations include poor growth, gastrointestinal signs and symptoms, iron deficiency anemia and pulmonary hemosiderosis (PH). Precipitins to cow's milk (CM) were also considered a useful aid in recognizing hypersensitivity to CM [1]. The literature concerning HS is restricted to a few case reports or series, although the disease has always been described at infant or pediatric age [2]. The definition of the disease is lacking and the diagnosis is often delayed, since its presentation is uncommon with dissimilar manifestations. In the present review we aimed at presenting clinical, diagnostic and therapeutic characteristics of HS, starting from current evidence.

2. Search Methodology and Results

We carried out a non-systematic review including the most relevant studies on "Heiner Syndrome" (HS) present on databases including PubMed (https://www.ncbi.nlm.nih.gov/pubmed/ accessed on 26 March 2021), MEDLINE, The Cochrane Library, from their inception to 26 March 2021. The searched terms were "Heiner Syndrome" [all fields]; "pulmonary hemosiderosis" and "children" [all fields]; "pulmonary hemosiderosis" and "cow's milk" [all fields]; "pulmonary hemosiderosis" and "hypersensitivity" [all fields]. We found 16 studies. They were all clinical cases or consecutive case series, involving an overall pool of 61 patients. Findings were summarized narratively below for each study as well as in Table 1.

In order to better stratify the level of evidence for the diagnosis, we are herein proposing the first structured diagnostic criteria for HS to our best knowledge. This diagnostic approach consists of the following criteria:

(A)   Pulmonary symptoms and XR infiltrates or pulmonary hemosiderosis (PH);
(B)   Resolution after milk removal;
(C)   Recurrence after milk reintroduction.

HS diagnosis (HSD): (A) + (B) = probable disease; (A) + (B) + (C) = convincing disease (Figure 1).

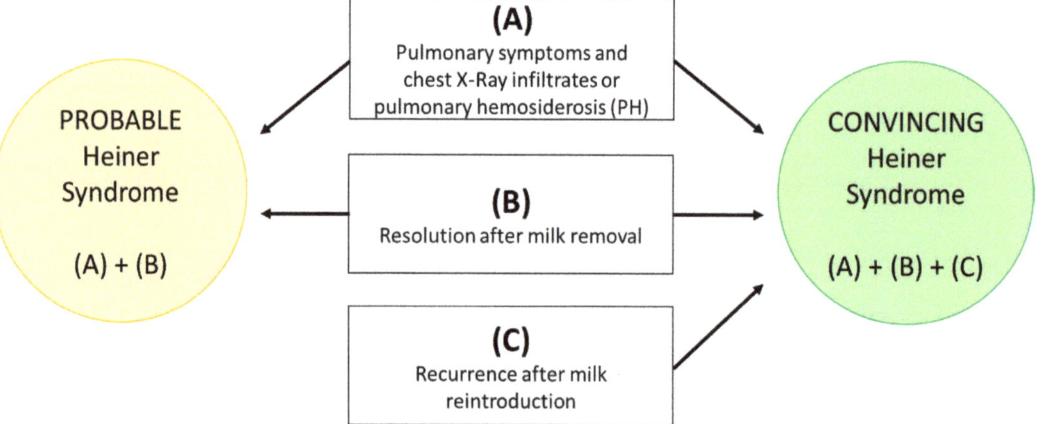

Figure 1. Proposed approach for Heiner syndrome diagnosis.

Table 1. Characteristics of the included studies of Heiner syndrome.

	Authors, Year	Country	n. Cases	Age at Onset (Months)	Signs/Symptoms	Lung Radiologic Infiltrates	Pulmonary Hemosiderosis	Hemosiderosis Diagnosis	Milk Precipitins	Delayed Hyper-sensitivity	Allergic Sensitization to Milk [sIgE and/or skin prick test (SPT)]	Improvement Upon Milk Avoidance	Recurrence Upon Milk Reintroduction
1	Heiner, 1960	USA	7	1–17	chronic cough (7/7); wheezing (7/7); chronic rhinitis (7/7); frequent fever (7/7); frequent earache (6/7); diarrhea (6/7); vomiting (5/7); hemoptysis (4/7)	yes, recurrent	yes (5/7)	gastric or bronchial aspirates (4/7); pulmonary needle biopsy (1/7)	yes	intradermal test (ID) late response pos (4/7)	ID +ve (7/7)	yes (6/6)	yes (2/6)
2	Holland, 1962	USA	22	4–12	respiratory disease; failure to thrive (FTT); anemia; splenomegaly; hepatomegaly	non specified (NS)	NS	NS	yes	NS	NS	yes (22/24 on milk-free diet)	NS
3	Chang, 1969	USA	1	9	FTT ; anemia; chronic recurrent lung disease	yes	NS	NS	yes	NS	NS	yes	doubtful, based on clinical history (CH)
4	Archer, 1971	England	1	13	iron-deficiency anemia; lethargy; pallor; bloody vomit; severe cardiac failure; hemoptysis	yes	yes	needle biopsy	no	NS	SPT –ve	yes	doubtful, based on CH
5	Boat, 1975	USA	6	7–48	idiopathic chronic or recurrent pulmonary disease; upper respiratory symptoms; FTT (3/6); frequent regurgitation and watery stools (1/6). SOF (1/6): Iron deficiency (5/6); Anemia (4/6); Eosinophilia (4/6); right ventricular hypertrophy (3/6); adenoid hypertrophy (3/6)	yes (6/6)	yes (5/6)	gastric washing or bronchoalveolar lavage (BAL)	yes (6/6)	4 ID delayed response	total IgE –ve (6/6); SPT +ve (5/6)	yes (5/6); 1/6 loss of data	yes (1/6)
6	Stafford, 1977	USA	7	8–48	wheezing (5/9); chronic rhinitis (3/9); large adenoids/tonsils (4/9); anemia (4/9); gastrointestinal symptoms (4/9); eosinophilia (6/9)	yes (7/9)	yes (3/9)	gastric washing and BAL	yes (9/9)	lymphocyte response (3/3)	SPT +ve (6/9), sIgE +ve (5/8)	NS	NS
7	Fossati, 1992	Italy	1	7 years	anemia; respiratory symptoms	yes	yes	NS	yes	NS	NS	yes	NS
8	Torres, 1996	Spain	1	0 (5 days)	vomiting with blood; respiratory failure; restrictive miocardiopathy; anemia; eosinophilia	yes (chest X-Ray, CXR)	yes	BAL	NS	NS	neg	yes	symptoms not reported
9	Moissidis, 2005	USA	8	4–29	cough (7/9); wheezing (3/9); dyspnea (1/9); hemoptysis (2/9); nasal congestion (3/9); recurrent otitis media (OM) (3/9); recurrent fever (4/9); gastrointestinal symptoms (5/9); in 7/8; hematochezia (1/9); FTT (2/9); eosinophilia in 5/8	yes (9/9)	yes (1/9)	NS	yes (6/6)	NS	sIgE (1/3), SPT (1/2)	yes (8/9)	positive challenge (3/3)
10	Sigua, 2013	USA	1	12	persistent cough; progressive anorexia; intermittent fever; weight loss; iron deficiency anemia	yes (CXR)	no	BAL	yes	NS	sIgE –ve	yes	doubtful, based on CH
11	Yavuz, 2014	Turkey	1	36	respiratory distress; hemoptysis; recurrent bronchitis; FTT; iron deficiency anemia; eosinophilia; increased inflammatory index;	yes (CXR, CT)	yes	BAL	NS	NS	sIgE +ve	yes	doubtful, based on CH
12	Mourad, 2015	USA	1	17	severe anemia; respiratory distress	yes (CXR)	yes	BAL	IgG	NS	sIgE +ve	yes	doubtful, based on CH
13	Alsukhon, 2017	USA	1	2	FTT; recurrent diarrhea; persistent cough; tachypnea; high inflammatory markers	yes (CXR)	NS	NS	IgG4	NS	sIgE –ve	yes	NS
14	Ojuawo, 2019	Nigeria	1	4	FTT; cough; dyspnea; wheeze; rhinitis; gastrointestinal symptoms; anemia	yes (CXR)	NS	NS	NS	NS	NS	yes	NS doubtful, based on CH
15	Koc, 2019	Turkey	1	6	massive hemoptysis; hematemesis; deep anemia	yes (CXR, CT)	yes	gastric washing	NS	NS	NS	yes	NS
16	Liu, 2020	China	1	4	respiratory failure; hematochezia; diarrhea; elevated WBC and C-reactive protein	yes (CXR, CT)	no	sputum or fasting gastric fluid	NS	NS	–ve	yes	NS

Abbreviations: BAL, bronchoalveolar lavage; CH, clinical history; CT, computerized tomography; CXR, chest X-Ray; ID, intradermal test; NS, non specified; SPT, skin prick test; –ve, negative; +ve, positive.

Milk precipitins have been evaluated only in some studies, mainly in the oldest ones. Based on current data and specifically due to the heterogeneity in the methodologies applied for laboratory tests, we decided to exclude laboratory parameters from the diagnostic approach.

(1) **Heiner et al., 1960** [3]—In 1960, Heiner [3] (from who the name of the syndrome originates) first reported a chronic respiratory disease associated with multiple CM precipitins in the sera of seven children aged 6 weeks to 17 months. All the patients presented a chronic respiratory disease with maximal severity of clinical manifestations at the time of introduction of raw CM in the diet associated with the other signs and symptoms that define the syndrome, mostly iron deficiency anemia, gastrointestinal signs and symptoms, poor growth and PH documented by gastric or bronchial aspirates. Six out of seven patients overcame their disease by changing the milk content (i.e., milk processing or exclusion of milk proteins) in their diet, by using evaporated or boiled milk (n = 2), extensively hydrolyzed casein formula (n = 1) or soymilk (n = 3). Overall, the milk protein avoidance resulted in the complete disappearance of clinical manifestations and remission of the hematologic status. One patient spontaneously overcame the disease without dietary restriction between 2.5 to 3.5 years (y) of age. When milk was reintroduced after avoidance, two out of six showed a clinical and imaging relapse, and four became tolerant or partially tolerant to milk after intervals without clinical manifestations on a restricted diet ranging from three to six months in duration. **Comment:** The first description of HS. **HSD:** Convincing in two cases; probable in four patients.

(2) **Holland et al., 1962** [4]—Stimulated by Heiner's observations, Holland et al. examined serum specimens from 1618 infants and children with the same technique, finding precipitins in 87 of them [4]. Patients of this population showed signs and symptoms suggestive of the syndrome but also different clinical manifestations, such as isolated upper respiratory diseases, hepatosplenomegaly and congenital heart diseases. Only 17 patients were reported to show respiratory signs and symptoms. No data on X-rays were reported. Because of the heterogeneity of the clinical manifestations and the limited number of diagnostic exams in this population, we selected 22/87 patients who improved during the CM diet period. No attempt of reintroduction was performed. **Comment:** We extrapolated 22 cases with suspected HS from a large and heterogeneous cohort. **HSD:** Nobody with probable or convincing clinical criteria.

(3) **Chang et al., 1969** [5]—Chang described the clinical case of a 9-month-old girl admitted to the hospital for failure to thrive (FTT), anemia and chronic recurrent lung disease starting in the first weeks of age [5]. She underwent a lengthy diagnostic process, until the finding of milk precipitins suggested an HS diagnosis. The patient was then placed on a soymilk diet with clinical resolution. Flare-up signs and symptoms and radiological relapse due to the poor adherence to the diet are mentioned. However, controlled milk reintroduction was not performed. **Comment:** Single case report. No detailed data are reported about the follow-up. **HSD:** Probable.

(4) **Archer, 1971** [6]—Archer reported the clinical case of a 13-month-old girl with a severe heart failure based on a profound iron deficiency anemia and idiopathic PH, diagnosed by needle biopsy [6]. All of the immunological tests performed in order to investigate a CM sensitization, including serum precipitins, skin prick test (SPT) and immunoglobulin, were negative. Notwithstanding, a milk-free diet was commenced with a good clinical and radiological response. The first relapse during her first week with a childminder was probably due to the inadvertent administration of CM. **Comment:** The girl was admitted one year before with some symptoms and treated successfully with antibiotics. Results of CM reintroduction doubtful. **HSD:** Probable.

(5) **Boat et al., 1975** [7]—In this study, 6 children with high titers of milk precipitins were identified by screening the sera of 160 children with idiopathic chronic lung disease associated with typical manifestations of milk-induced PH [7]. Elimination of CM

from the diet, symptomatic therapy and adenoidectomy (when indicated) resulted in improvement in six out of six patients. **Comment:** Even if six out of six patients recovered in 5–21 days after milk removal, only one was rechallenged (Patient C) and one (Patient B) developed pneumonia within six months upon CM reintroduction. **HSD:** Convincing in one case; probable in five patients.

(6) **Stafford et al., 1977** [8]—Nine patients with respiratory signs and symptoms and milk precipitins were enrolled in this study in order to elucidate the immunopathologic mechanisms involved in milk-induced PH. No demonstration of a unique immunologic mechanism associated with milk-induced PH in the patients studied [8]. No clinical data about CM withdrawal were reported. **Comment:** Focus on immunological patho-mechanisms with poor clinical description of the enrolled participants. **HSD:** Nobody with probable or convincing criteria.

(7) **Fossati, 1992** [9]—A 7-year-old girl was admitted to hospital because of anemia, and a PH was diagnosed [9]. Precipitating antibodies were also found. A marked improvement of clinical manifestations and XR results were found after removal of CM from the diet. **Comment:** Single clinical case report. No data on reintroduction. **HSD:** Probable.

(8) **Torres et al., 1996** [10]—This study provides an interesting immunological overview on the HS based on data from a single clinical case of a girl [10]. Specifically, the authors speculate that an inflammatory response occurred after CM intake in the presented clinical case. A 5-day-old female newborn was admitted in an emergency department because of a severe bradycardia due to a myocarditis associated with assessed PH and anemia. Although the total specific IgE and specific IgG to milk proteins were below the detection limits, the patient underwent two oral food challenges (OFC). During the in vivo tests, hematocrit, histamine, tryptase and ECP (eosinophil cationic protein) in blood and BAL, and N-methyl-histamine (NMH) in urine were measured before and at multiple times during the administration of standard formula (first OFC) and non-milk enteral nutrition (second OFC). When the girl was fed with CM, a remarkable increment of all the tested inflammatory mediators was reported; conversely, the hemoglobin level dropped significantly. Then, the same OFC was done with non-milk enteral nutrition with any variation being registered. These data are unfortunately not supported by clinical and radiologic information. Challenge. **Comment:** Unique case report of PH in which milk OFC induced an increase of inflammatory mediators suggesting a T cell-mediated pathogenesis of HS. **HSD:** Probable.

(9) **Moissidis et al., 2005** [11]—Moissidis et al. reviewed eight cases of children affected by upper respiratory tract symptoms [11]. All cases presented radiological imaging with pulmonary infiltrates, and one had HP (defined as iron-laden macrophages in the bronchoalveolar lavage, gastric washing and open lung biopsy). Seven out of eight had gastrointestinal symptoms. High titers of precipitating antibodies to CM proteins were demonstrated in six out of six patients studied. However, HS was confirmed by the improvement of the clinical and radiological findings after a CM-free diet and relapse when a reintroduction was attempted in three out of six cases. **Comment:** The most detailed paper on the topic. However, cases were evaluated at different times and under different circumstances; therefore, specific data were not available for each patient. **HSD:** Convincing in three cases; probable in five patients.

(10) **Sigua et al., 2013** [12]—A 12-month-old boy with multifocal pneumonia that was refractory to protracted antibiotic treatment was suspected to suffer from HS [12]. The clinical history showed that the boy underwent a milk-free diet from the first to the tenth month of age because of suspected non-IgE-mediated CM non-bloody diarrhea. HS appeared at CM reintroduction. Serum-precipitating IgG antibodies to all nine CM protein fractions tested were strongly positive. He underwent a strict soy-based diet from 12 months of age with prompt clinical remission and complete resolution of the previously identified pulmonary opacities at a chest X-ray performed at 14 months.

Comment: A single patient with HS after a previous history of non-IgE-mediated CM gastrointestinal symptoms. **HSD:** Probable.

(11) **Yavuz et al., 2014** [13]—A 3-year-old boy was referred to the emergency service with respiratory distress and hemoptysis [13]. Because of iron deficiency anemia, a BAL cytological examination was performed in order to confirm a PH. Precipitins were not determined. The patient overcame the disease through a CM avoidance diet. However, a low dose of both prednisolone and azathioprine was also prescribed. Furthermore, authors described that the patient in the next five years had many relapses because of failure to receive the prescribed medications and poor adherence to the diet. Moreover, during a hemoptysis attack, he showed new symptoms, such as edema, hematuria and hypertension. On this occasion, rapidly progressive glomerulonephritis was diagnosed on the basis of the histopathological findings and treated with a combination of cyclophosphamide and methylprednisolone. **Comment:** In this case report, an elimination diet and drugs were administrated together for an extended period of time, and during the follow-up the compliance was scarce. Therefore, it is difficult to differentiate the effects of each treatment and the actual cause-effect relationship. **HSD:** Probable.

(12) **Mourad et al., 2015** [14]—The clinical case of a 17-month-old boy with idiopathic PH was described by Mourad et al. [14]. BAL demonstrated an abundance of fresh red blood cells and iron-laden macrophages. The CM-specific IgE level was only slightly elevated (1.42 IU/mL). IgG antibody levels to CM proteins were markedly elevated. In spite of the severity of the clinical conditions (i.e., severe anemia and respiratory failure with acidosis), the strict CM-free diet allowed the boy to overcome the disease. Hydrocortisone was also administrated, but it is not clear when it was introduced and for how long. A relapse was reported because the mother, while on raw CM avoidance, started feeding the patient with baked CM products. **Comment:** Single clinical case report. Hydrocortisone was also used. Controlled reintroduction not performed. **HSD:** Probable.

(13) **Alsukhon et al., 2017** [15]—A 2-month-old male with recurrent diarrhea and FTT had persistent cough, tachypnea and high inflammatory markers despite antibiotic therapy for pneumonia [15]. An amino acid-based formula gave improvement in inflammation and respiratory function. **Comment:** Single case report. Milk reintroduction not performed. **HSD:** Probable.

(14) **Ojuawo et al., 2019** [16]—Ojawo et al. described the clinical case of a 16-week-old boy with FTT, dyspnea and anemia who acceded to the emergency department in Nigeria [16]. Neither antibiotic treatment nor sodium citrate, administered for the suspicion of a renal tubular acidosis, modified his condition. Diagnosis of HS was based on the constellation of clinical features, XR results, and subsequent resolution after stopping CM. Parents on a follow-up visit reported occasional cough and rhinitis whenever CM was reintroduced. **Comment:** Single case report. No controlled CM reintroduction reported. **HSD:** Probable.

(15) **Koc et al., 2019** [17]—A 6-month-old infant with massive hemoptysis, hematemesis and deep anemia was treated for bronchopneumonia four times [17]. When he was admitted to the emergency department, both chest-X ray and computerized tomography documented many lung opacities and hemosiderin-laden macrophages were found in the patient's fasting stomach fluid examination, confirming the diagnosis of PH. The boy was discharged with a CM-free diet, with complete clinical and radiological recovery. **Comment:** No laboratory data were reported, and no milk reintroduction test was reported. **HSD:** Probable.

(16) **Liu et al., 2020** [18]—Liu et al. described a 4-month-old boy with a chronic pulmonary syndrome whose main presenting symptom was a persistent hematochezia since the tenth day of life [18]. Gastrointestinal endoscopic biopsy showed granulation tissue infiltrated by acute and chronic inflammatory cells, including some eosinophils. Additionally, in this case, the improvement of both clinical and radiologic findings

after the elimination of milk suggested the diagnosis of HS. In addition to the CM elimination diet, the patient was treated with methylprednisolone (1 mg/kg) and montelukast. **Comment:** Single case report. No milk reintroduction test reported. **HSD:** Probable.

3. Discussion

To the best of our knowledge, the present article represents the first review on this rare disease. The data shown suggest a critical approach to the disorder (Table 1).

3.1. Age at Onset

The clinical onset of the disease has been described typically by the age of 1 month to 48 months, but it can develop even during the first 5 days of life, as reported by Torres et al. [10]. However, it can also appear later (the oldest patient was 5 years old) [4]. Nevertheless, there was a frequent delay in diagnosing this disease, due to its various modes of presentation and lack of standardized diagnostic criteria. The past medical history of affected children was always unremarkable. A family history of allergic disorders was often present. Differently from immediate-onset IgE-mediated CM allergy, HS did not display signs and symptoms before several days or weeks after CM consumption.

3.2. Etiology

Although HS is more likely to be induced by homogenized CM, the disease also may occur in some infants fed with CM-derived formula. Furthermore, it has been speculated that it can be related to non-IgE-mediated allergy to food proteins differently from CM at an older age (e.g., soy, egg, pork, wheat and peanut) [14,19]. In this context, a single case of PH hemosiderosis due to buckwheat has also been reported [20].

3.3. Clinical Characteristics

Respiratory features of the disease included persistent cough, dyspnea, tachypnea, wheezing, occasional sputum production and rales. The peculiarity of pneumonia in these case series was the refractoriness of antibiotic treatments. Of note, in most cases, the additional administration of anti-inflammatory drugs probably might have resolved hypersensitivity pneumonia or idiopathic PH (IPH). The most commonly described systemic clinical manifestations were intermittent fever, progressive anorexia and FTT. Inflammatory markers were usually found to be high. Eosinophilia and severe iron deficiency anemia were frequently described at blood count examination. Gastrointestinal manifestations were reported in about half of the patients and included frequent vomiting or diarrhea. Rarely, lymph node hypertrophy with hepatomegaly, splenomegaly and hypertrophied tonsils or adenoids were labeled [7]. Noticeably, lymphonodular hyperplasia in biopsy was found in a child with HS-manifesting hematochezia [18].

Clinically, the disease can be complicated with cardiopulmonary involvement, such as alveolar hypoventilation, massive acute PH, pulmonary hypertension and *cor pulmonale*, or nephrological ones, such as crescentic glomerulonephritis [8,13]. These characteristics contributed drastically to morbidity and emerged in situations of overdue diagnosis and management. In particular, a delayed manifestation of the disease is episodic hemoptysis, which may represent a PH with repeated episodes of intra-alveolar bleeding, hemosiderin deposition in alveolar macrophages, followed by the development of pulmonary fibrosis and severe anemia [8]. PH may occur as a primary disease of the lung (also called IPH) or secondary to cardiac diseases, bleeding disorders, collagen-vascular diseases or systemic vasculitis. IPH, if not treated, leads to progressive pulmonary fibrosis and may be lethal [21].

3.4. Pathogenesis and Immunological Implications

The exact mechanism that triggers HS is not fully understood. Feasibly, the formation of immune complexes is strongly suspected (Gell and Coombs type III reaction) and the cell-mediated reaction (Gell and Coombs type IV reaction) may contribute to the development of this challenging disease.

Some cases showed positive skin tests [3,7], high serum total IgE levels [7], high milk-specific IgE antibodies [13,14] or circulating immune complexes [3]. A significant increase of histamine and ECP in BAL several hours after a milk OFC was reported [10]. In one case report, deposits of immunoglobulins, complement, fibrin and milk protein antigens diffusely scattered were described on immunofluorescence studies of lung tissue biopsies [2]. It is probable that a cause concurring to HS is the aspiration of milk, in particular among patients with an uncoordinated swallowing mechanism, tracheal/esophageal anomaly or gastroesophageal reflux. However, in the paper of Boat et al., this condition was ruled out [7]. Concerning the data on delayed immunity, in some cases a delayed skin test response to intradermal test [3,7] or a lymphocyte response [8] was reported. Other authors postulated that milk antigens might trigger an immune complex reaction resulting in multiorgan abnormalities, such as pulmonary, gastrointestinal and renal ones. In fact, pulmonary and gastrointestinal signs and symptoms were frequently associated [11] and granular immuno-deposits have been demonstrated along the glomerular basement membrane in a child with crescentic glomerulonephritis and PH [13].

Most studies, mainly the oldest ones, characteristically found high titers of precipitins (likely immunoglobulins of class G) against bovine milk proteins in the patients' sera, by using the Ouchterlony double-immunodiffusion technique [4,5,7,11]. However, it is not sufficiently explicable why some children develop precipitating antibodies to ingested protein and other children do not. Moreover, it is not known if these precipitins play a causative role in the disease. Children with precipitins usually have an increased incidence of recurrent respiratory tract diseases, anemia and hepatosplenomegalia. However, precipitating IgG antibodies to milk are not pathognomonic of the disease, since they have been found among around 1% of healthy children in the absence of clinical manifestations [4], and in 4% to 6% of children with chronic disorders, including celiac disease, cystic fibrosis, IgA deficiency, Down's syndrome, Wisckott–Aldrich syndrome and Hurler's syndrome [7]. Additionally, methods used for detecting precipitins are obsolete and reports on them are timeworn. Therefore, the role of these antibodies should be critically considered.

In other more recent reports, high values of specific IgG to milk proteins were found using an immunoenzimatic technique [14]. In one case report [15] CM IgG4 was found to be elevated. Even in these cases, the role of this specific CM IgG is not clear.

3.5. Pulmonary Modifications and Immunofluorescence Studies

Chest roentgenograms displayed variable patchy and transient infiltrates, frequently associated with areas of atelectasis, consolidation, reticular opacities, pleural thickening or hilar lymphadenopathy. The lung biopsy obtained in patients who had hemoptysis showed an abnormal accumulation of hemosiderin in the lungs, which resulted in alveolar hemorrhage or PH [2]. Among patients with anemia and hemoptysis, PH was verified by the demonstration of iron-laden macrophages by using Prussian Blue staining of bronchial aspirates or morning gastric washes [8].

3.6. Diagnostic Criteria

As described in detail above, in order to better stratify the level of evidence for the diagnosis, we are herein proposing the first structured diagnostic approach for the diagnosis of HS to the best of our knowledge. We auspicate that this approach may allow clinicians to stratify patients with a clinical history consistent with the suspicion of HS in probable (criteria A + B) or convincing (criteria A + B + C) HS (Figure 1). According to our criteria, only 6 out of 61 patients had a convincing clinical diagnosis of HS, in 25 patients the clinical diagnosis was probable and in the others the HS diagnosis was doubtful (Table 1). Due to the heterogeneity in the methodologies applied for laboratory tests and missing reporting/lack of data, we decided to exclude laboratory parameters from the diagnostic criteria. We suggest in the future that milk-specific IgG tests with current diagnostic methods (i.e., immunoenzymatic), in the case of suspected HS, could be studied. Nevertheless, further points remain questionable in our proposed diagnostic approach. First, some cases of occurrence of signs and symptoms during exclusive maternal breastfeeding were reported (e.g., in three out of seven cases in Heiner et al. [3], leading to the question of whether or not minimal quantities of CM passing in breast milk are capable of inducing a clinical response through an IgG-mediated mechanism in the infant). As a second observation, one case of resolution of signs and symptoms without dietary restriction has been described (one out of seven patients from Heiner et al.). Again, a real, controlled OFC of milk was reported only in 6 cases [3,7,11] and a recurrence of clinical manifestations was reported in a further 6 cases (although without details on exact timing of the symptoms' onset) based only on clinical history [5–7,13,14,16], for a total of 12 cases. Moreover, in some cases [13,14] pharmacological therapy was associated to the CM elimination diet, making it difficult to differentiate the effect of each single treatment. Spontaneous resolution of signs and symptoms also occurred in some cases of HS [1]. Furthermore, some studies did not report on the follow-up and specifically on the outcome of any CM reintroduction. In conclusion, even if in a few cases a convincing diagnosis can be made using specific criteria, certainty is lacking due to the incomplete clinical and imaging monitoring of the OFC and the lack of control cases.

3.7. Differential Diagnosis

In the differential diagnosis (DD) bronchial asthma, chronic aspiration, acute and chronic lower respiratory tract infections, including fungal ones, cystic fibrosis, foreign body, hypersensitivity pneumonitis, bronchopulmonary aspergillosis, secondary PH and IPH should be considered [11]. Cystic fibrosis essentially can be excluded by a normal chloride level on the sweat test and second line tests if there is a strong suspicion. Infectious causes should be ruled out with tuberculin skin test and available microbial tests, and empiric antibiotic treatments characteristically are inefficacious. Thereafter, a vasculitis or autoimmune disorder can be considered. The characteristic pulmonary hemorrhage attacks cannot be enlightened by modest bronchial asthma. The lack of chronic inhalant exposure and BAL examination can rule out hypersensitivity pneumonitis. Bronchopulmonary aspergillosis is unlikely if skin reactivity to *Aspergillus* antigens is negative or precipitating serum antibodies to *A. fumigatus* are absent. No hemorrhagic focus nor foreign bodies can be found on bronchoscopy. In particular, DD should consider IPH, that includes the classic triad hemoptysis, radiologic lung infiltrate and iron deficiency anemia, having a more severe course and prognosis [22]. Moreover, IPH occurs in older children, and it is rarely associated with gastrointestinal symptoms [23] (Table 2).

Table 2. Differential diagnosis between Heiner syndrome (HS) and idiopathic pulmonary hemosiderosis (IPH).

	HS	IPH
Age	infants or young children	older children/adults
Hemosiderosis	often	always
GI symptoms	often	rarely
Precipitins	yes	no
Response to diet	yes	no
Prognosis	good	variable

3.8. Natural History

The disease's effects were reversible by stopping CM consumption. In fact, signs and symptoms could last within a time range of 5 to 21 days after the CM withdrawal. Before CM reintroduction, some children showed spontaneous tolerance to pasteurized or boiled CM [4]. Reoccurrence of clinical manifestations was reported with CM reintroduction [3,5,6,10,11,14]. However, it is believed that patients usually definitively outgrow this hypersensitivity, and they can tolerate CM within a few years [11,24,25].

3.9. Treatment

Clinical improvement with strict elimination of CM proteins sustains the diagnosis of HS. Infants may be fed by a milk substitute, such as extensively hydrolyzed protein formula, soy-based formula or synthesized free amino acid formula. Improvement of signs and symptoms occurs in few days and X-ray images in weeks. When a confirmatory CM reintroduction was performed [11], respiratory clinical manifestations also occurred in days or weeks. An early exclusion of the triggering food from the diet is crucial, since chronic PH induces pulmonary fibrosis which can be fatal [6]. However, recovery may occur also without exclusion of the culprit food (e.g., 1/7 in Heiner's report [3]).

Even if the most striking criterion of HS is the dramatic response to the exclusion diet, initially, in some cases, appropriate treatment, e.g., bronchodilators, antihistamines, systemic or inhaled steroids and iron, may be needed. A short cycle of oral corticosteroids remains the first-line therapy for acute attacks. In more severe cases of HS, other immunomodulatory treatments may be helpful, such as hydroxychloroquine, azathioprine or cyclophosphamide. On the contrary, antibiotic therapy seems not to be useful.

4. Conclusions

Although HS has been described as a hypersensitivity disease due to CM, there are still pros and cons about its real existence (Table 3), and a certainty of diagnosis is lacking. A challenge test was performed in a few cases and always in an open manner. Again, signs and symptoms develop in the hours or days after milk consumption, and they disappear after the elimination diet. Precipitating IgG antibodies are an old-fashioned laboratory test reported in some patients during the disease; precipitins diminish or disappear during the elimination diet. However, they are not pathognomonic nor specific for the disease diagnosis and their pathogenetic role is still unclear. The prognosis is generally good, even if the extent of the exclusion diet necessary to reach a complete recovery is unclear. HS is a very intriguing disease, in some ways still controversial, but it is important to know and suspect this rare syndrome in any infant or young child with unexplained chronic pulmonary clinical manifestations. The diagnosis should be proved by clinical and radiologic improvements after strict CM avoidance with the recurrence of signs of symptoms and imaging features after a controlled CM reintroduction.

Table 3. The real existence of Heiner syndrome: pros and cons.

Pros	Cons
Multiorgan involvement (in particular lung and GI)	Absence of case–control studies
Detection of precipitating antibodies	Precipitating antibodies not pathognomonic
Scarce response to non anti-inflammatory drugs	In most cases the additional administration of anti-inflammatory drugs probably resolved hypersensitivity pneumonia or PH
Clinical improvement after milk removal	The presence of milk in pulmonary infiltrates reported only in one case
Symptoms' reoccurrence after milk reintroduction	Confirmatory challenge not provided in most cases and/or not adequately performed

Considering the high clinical impact of the disease and the associated morbidity, more attention should be devoted to it, both in terms of clinical suspicion and research on the underlining patho-mechanisms with the detection of reliable biomarkers. We highlight the need for more stringent diagnostic criteria that combine both clinical manifestations and imaging features. Moreover, a follow-up evaluation with a well-designed CM OFC/regular reintroduction is of paramount importance to better understand this disease in terms of prognosis and duration. In summary, the future establishment of validated diagnostic criteria, the awareness of specific clinical manifestations and specific imaging features and the results of CM OFC will help health professionals in clinical practice to suspect the disease and to refer patients to the appropriate specialists.

Author Contributions: E.N. conceived the study, coordinated it, helped in drafting the manuscript and critically reviewed it. S.A. and C.M. participated in the study's design, carried out the literature research and wrote the first draft of the manuscript. L.P., F.M., M.G. and S.B. reviewed and provided feedback. All authors have read and agreed to the published version of the manuscript.

Funding: This research received no external funding.

Institutional Review Board Statement: Not applicable.

Informed Consent Statement: Not applicable.

Data Availability Statement: Not applicable.

Acknowledgments: Not applicable.

Conflicts of Interest: The authors declare no conflict of interest.

References

1. Heiner, D.C.; Sears, J.W.; Kniker, W.T. Multiple precipitins to cow's milk in chronic respiratory disease. A syndrome including poor growth, gastrointestinal symptoms, evidence of allergy, iron deficiency anemia, and pulmonary hemosiderosis. *Am. J. Dis. Child* **1962**, *103*, 634–654. [CrossRef] [PubMed]
2. Lee, S.K.; Kniker, W.T.; Cook, C.D.; Heiner, D.C. Cow's milk-induced pulmonary disease in children. *Adv. Pediatr.* **1978**, *25*, 39–57.
3. Heiner, D.C.; Sears, J.W. Chronic respiratory disease associated with multiple circulating precipitins to cow's milk. *Am. J. Dis. Child.* **1960**, *100*, 500–502.
4. Holland, N.H.; Hong, R.; Davis, N.C.; West, C.D. Significance of precipitating antibodies to milk proteins in the serum of infants and children. *J. Pediatr.* **1962**, *61*, 181–195. [CrossRef]
5. Chang, C.H.; Wittig, H.J. Heiner's syndrome. *Radiology* **1969**, *92*, 507–508. [CrossRef]
6. Archer, J. Idiopathic Pulmonary Hemosiderosis Treated with a Milk-free Diet. *Proc. Roy. Soc. Med.* **1971**, *64*, 1217–1218. [CrossRef] [PubMed]
7. Boat, T.F.; Polmar, S.H.; Whitman, V.; Kleinerman, J.I.; Stern, R.C.; Doeshuk, C.F. Hyperactivity to cow milk in young children with pulmonary hemosiderosis and cor pulmonale secondary to nasopharyngeal obstruction. *J. Pediatr.* **1975**, *87*, 23–29. [CrossRef]
8. Stafford, H.A.; Polmar, S.H.; Boat, T.F. Immunologic Studies in Cow's Milk-induced Pulmonary Hemosiderosis. *Pediatr. Res.* **1977**, *11*, 898–903. [CrossRef]
9. Fossati, G.; Perri, M.; Careddu, G.; Mirra, N.; Carnelli, V. Pulmonary hemosiderosis induced by cow's milk proteins: A discussion of a clinical case. *Pediatr. Medica Chir. Med Surg. Pediatrics* **1992**, *14*, 203–207.
10. Torres, M.J.; Girón, M.D.; Corzo, J.L.; Rodriguez, F.; Moreno, F.; Perez, E.; Blanca, M.; Martinez-Valverde, A. Release of inflammatory mediators after cow's milk intake in a newborn with idiopathic pulmonary hemosiderosis. *J. Allergy Clin. Immunol.* **1996**, *98*, 1120–1123. [CrossRef]
11. Moissidis, I.; Chaidaroon, D.; Vichyanond, P.; Bahna, S.L. Milk-induced pulmonary disease in infants (Heiner syndrome). *Pediatr. Allergy Immunol.* **2005**, *16*, 545–552. [CrossRef]
12. Sigua, J.A.; Zacharisen, M. Heiner syndrome mimicking an immune deficiency. *WMJ Off. Publ. State Med Soc. Wis.* **2013**, *112*, 215–217.
13. Yavuz, S.; Karabay-Bayazıt, A.; Yılmaz, M.; Gönlüşen, G.; Anarat, A. Crescentic glomerulonephritis in a child with Heiner syndrome. *Turk. J. Pediatr.* **2014**, *56*, 661–664. [PubMed]
14. Mourad, A.A.; Parekh, H.; Bahna, S.L. A 17-month-old patient with severe anemia and respiratory distress. *Allergy Asthma Proc.* **2015**, *36*, 506–511. [CrossRef] [PubMed]
15. Alsukhon, J.; Leonov, A.; Elisa, A.; Koon, G. P327 Food-induced pulmonary hemosiderosis. *Ann. Allergy Asthma Immunol.* **2017**, *119*, S77. [CrossRef]
16. Ojuawo, A.B.; Ojuawo, O.B.; Aladesanmi, A.O.; Adio, M.O.; Abdulkadir, M.B.; Mokuolu, O.A. Heiner Syndrome: An uncommon cause of failure to thrive. *Malawi Med. J.* **2019**, *31*, 227–229. [CrossRef]
17. Koc, A.S.; Sucu, A.; Celik, U. A different clinical presentation of Heiner syndrome: The case of diffuse alveolar hemorrhage causing massive hemoptysis and hematemesis. *Respir. Med. Case Rep.* **2019**, *26*, 206–208. [CrossRef]
18. Liu, X.-Y.; Huang, X.-R.; Zhang, J.-W.; Xiao, Y.-M.; Zhang, T. Hematochezia in a Child with Heiner Syndrome. *Front. Pediatr.* **2020**, *7*, 551. [CrossRef] [PubMed]
19. Heiner, D.C.; University of California Los Angeles, Torrance, CA, USA. Personal Verbal Communication, 2015.
20. Agata, H.; Kondo, N.; Fukutomi, O.; Takemura, M.; Tashita, H.; Kobayashi, Y.; Shinoda, S.; Nishida, T.; Shinbara, M.; Orii, T. Pulmonary hemosiderosis with hypersensitivity to buckwheat. *Ann. Allergy Asthma Immunol.* **1997**, *78*, 233–237. [PubMed]
21. Soergel, K.H.; Sommers, S.C. Idiopathic pulmonary haemosiderosis and related syndromes. *Am. J. Med.* **1962**, *32*, 499–511. [CrossRef]
22. Saha, B.K. Idiopathic pulmonary hemosiderosis: A state of the art review. *Respir. Med.* **2021**, *176*, 106234. [CrossRef] [PubMed]
23. Taytard, J.; Nathan, N.; De Blic, J.; Fayon, M.; Epaud, R.; Deschildre, A.; Troussier, F.; Lubrano, M.; Chiron, R.; Reix, P.; et al. New insights into pediatric idiopathic pulmonary hemosiderosis: The French RespiRare® cohort. *Orphanet J. Rare Dis.* **2013**, *8*, 161. [CrossRef]

24.   Lee, E.J.; Heiner, D.C. Allergy to Cow Milk. *Pediatrics Rev.* **1986**, *7*, 7. [CrossRef] [PubMed]
25.   Williams, S.; Craver, R.D. Cow's milk-induced pulmonary hemosiderosis. *J. State Med Soc. Off. Organ State Med Soc.* **1989**, *141*, 19–22.

Article

Nut Allergy: Clinical and Allergological Features in Italian Children

Sylvie Tagliati [1,†], Simona Barni [2,†], Mattia Giovannini [2,*], Giulia Liccioli [2], Lucrezia Sarti [2], Tatiana Alicandro [3], Erika Paladini [4], Giancarlo Perferi [5], Chiara Azzari [5], Elio Novembre [2] and Francesca Mori [2]

[1] Pediatric Unit, Sant'Anna University Hospital of Ferrara, 44124 Ferrara, Italy; sylvie.tagliati@edu.unife.it

[2] Allergy Unit, Department of Pediatrics, Meyer Children's University Hospital, 50139 Florence, Italy; simonabarni@hotmail.com (S.B.); giulialiccioli@gmail.com (G.L.); lucrezia.sarti@unifi.it (L.S.); elio.novembre@unifi.it (E.N.); francesca.mori@meyer.it (F.M.)

[3] Division of Allergy and Clinical Immunology, Dermatology Unit, Surgical, Medical and Dental Department of Morphological Science Related to Transplant, Oncology and Regenerative Medicine, University of Modena and Reggio Emilia, 41124 Modena, Italy; tatiana.alicandro@gmail.com

[4] Pediatric Respiratory and Allergy Unit, Women's and Children's Health Department, University of Padua, 35128 Padua, Italy; erikapaladini@yahoo.it

[5] Section of Pediatrics, Department of Health Sciences, University of Florence and Meyer Children's Hospital, 50139 Florence, Italy; giancarlo.perferi@meyer.it (G.P.); chiara.azzari@unifi.it (C.A.)

* Correspondence: mattiag88@hotmail.it; Tel.: +39-0555662472

† These authors contributed equally to this work.

Citation: Tagliati, S.; Barni, S.; Giovannini, M.; Liccioli, G.; Sarti, L.; Alicandro, T.; Paladini, E.; Perferi, G.; Azzari, C.; Novembre, E.; et al. Nut Allergy: Clinical and Allergological Features in Italian Children. *Nutrients* **2021**, *13*, 4076. https://doi.org/10.3390/nu13114076

Academic Editor: Margarida Castell Escuer

Received: 13 September 2021
Accepted: 11 November 2021
Published: 15 November 2021

Publisher's Note: MDPI stays neutral with regard to jurisdictional claims in published maps and institutional affiliations.

Abstract: Background: Nut allergies are an increasingly frequent health issue in the pediatric population. Tree nuts (TN) and peanuts are the second cause of food anaphylaxis in Italy. Unfortunately, knowledge of the clinical characteristics of a TN allergy in Italian children is limited. Our study aimed to identify the clinical and allergological characteristics of Italian children with a nut allergy (TN and peanut). Methods: A retrospective observational analysis was performed on the clinical charts of children with a history of nut reaction referred to the allergy unit of the hospital from 2015 to 2019. The studied population was represented by children with a confirmed nut allergy based on positive prick by prick and/or serum-specific IgE to nut plus a positive nut oral food challenge. Demographic, clinical, and allergological features were studied and compared among different nuts. Results: In total, 318 clinical charts were reviewed. Nut allergy was confirmed in 113 patients. Most patients (85/113, 75%) had a familial history of allergy and/or a concomitant allergic disorder (77/113, 68%). Hazelnut and walnut were the more common culprit nuts observed in allergic children. Anaphylaxis was the first clinical manifestation of nut allergy in a high percentage of children (54/113, 48%). The mean age of the first nut reaction was statistically higher with pine nuts. Over 75% of children reported a single nut reaction. During the OFCs, the signs and symptoms involved mainly the gastrointestinal system (82/113, 73%) and resolved spontaneously in most cases. Severe reactions were not frequent (22/113, 19%). Conclusion: To our knowledge, this is the first Italian study that provided a comprehensive characterization of children with a nut allergy. These results are important for clinicians treating children with a nut allergy.

Keywords: children; nut allergy; oral food challenge; peanut; prick by prick; serum specific IgE; skin prick test; tree nut

1. Background

Nut allergies are an emerging health issue in the pediatric population [1], which are experiencing increasing prevalence in childhood and exhibiting important effects on the quality of life of children and their families [2–5]. Tree nuts (TN) and peanuts have been identified as the main culprits of fatal or near-fatal anaphylaxis, even with consumption in a small amount [6,7]. In Italy, TN and peanuts are the second-leading cause of food anaphylaxis and the first in North America [8,9]. TN include almonds, Brazil nuts, cashews, hazelnuts, macadamia nuts, pecan nuts, pine nuts, pistachios, and walnuts. On the contrary, peanuts are not considered as TN because they belong to the *Fabaceae* family and are classified as legumes [10]. In this study, for convenience, we used the term nut, which includes TN and peanuts.

Clinically, a nut allergy can present as a primary nut allergy or pollen food syndrome (PFS)/oral allergic syndrome (OAS). The primary nut allergy is usually characterized by systemic and severe reactions due to the presence of serum-specific IgE (s-IgE) against the major nut storage proteins (e.g., Ara h 2 for peanuts). Instead, PFS/OAS is usually characterized by mild and isolated signs and symptoms to the oropharynx. PFS/OAS manifests in patients with seasonal allergic rhinitis and a history of reaction to nuts due to the presence of s-IgE directly against heat-labile proteins (e.g., PR-10), homologous to those in pollen [8,11].

The diagnosis of a nut allergy is based on clinical history, prick by prick (PbP) results, and s-IgE detection [12,13]. Molecular allergen analysis is becoming a more utilized method and may improve accuracy for diagnosing [13]. The oral food challenges (OFCs) are still considered the gold standard for the diagnosing of nut allergies and are useful to distinguish between sensitization and a primary allergy [14].

The knowledge of clinical characteristics of nut allergies in Italy is markedly limited, especially in the pediatric population [15]. Hence, our study aimed to identify the demographic, clinical, and allergological characteristics of Italian children with different nut allergies, comparing these features between the various nuts.

2. Materials and Methods

We performed a retrospective observational analysis of the clinical charts of children with a history of nut reactions who were referred to the allergy unit of the hospital from January 2015 to December 2019. Written informed consent for all performed procedures was obtained from the children's parents. The code of the event report issued by the hospital is IR904-18-26854.

A skin prick test (SPT) for aeroallergens (including grass, artemisia, cypress, olive tree, hazel, birch, and poplar) and/or specific foods if the clinical history was suggestive of respiratory allergy (asthma and/or oculorhinitis) and food allergy was performed by using commercial extracts (Lofarma, Milan, Italy). When an SPT for specific foods was not available, we performed a PbP with fresh foods. Fresh nuts were used for PbPs, according to Ortolani et al. [16]. Both SPTs and PbPs were performed on the volar surface of the forearm with a lancet, as per the European standard [17]. The results were read after 15 min: a wheal diameter $\geq$ 3 mm was considered positive. Positive and negative controls were used—histamine (10 mg/mL; Lofarma, Milan, Italy) and normal saline, respectively.

In patients with a positive PbP to nuts, a s-IgE to nuts and the available molecular components (peanuts: (Ara h 1, Ara h 2, Ara h 3, Ara h 8, Ara h 9); hazelnuts: (Cor a 1, Cor a 8, Cor a 9, Cor a 14); walnuts: (Jug r 1, Jug r 3)) were detected by ImmunoCAP (Thermo Fisher Scientific, Uppsala, Sweden), following the manufacturer's instructions. A positive cut-off point was set at 0.1 kUA/L.

All the patients with positive tests in reference to the culprit nut underwent an OFC with the nut suspected as the cause for the reaction or beginning with the nut suspected as the cause for the first reaction in chronological order in the case of multiple reactions (according to clinical history and sensitization profile).

The OFC was performed under an allergist's supervision, and it was carried out according to international standards [18,19] adapted to the context of a one-day hospital setting [20]. The OFC was usually proposed to healthy children and postponed in case of acute diseases like fever, infectious gastroenteritis, or bronchitis. The protocol used for the nuts OFC is summarized in Table 1. The increasing doses were given every 20 min until completing the protocol or reaching the threshold dose for reaction. The OFC was considered positive if there were objective signs like urticaria, angioedema, vomiting, diarrhea, bronchospasm, hoarse voice, rhinitis, conjunctivitis, hypotension, or loss of consciousness within two hours after administration of the last food dose, which is the frame time of observation in the hospital setting for IgE-mediated food allergies. If there were reactions, patients were treated as needed and observed for a minimum of two hours until the clinical manifestations of the reaction resolved.

Table 1. Nut oral food challenge protocol.

	Dose (mg)	Almond (mg of Protein)	Cashew (mg of Protein)	Hazelnut (mg of Protein)	Peanut (mg of Protein)	Pine Nut (mg of Protein)	Pistachio (mg of Protein)	Walnut (mg of Protein)
	5	1.05	0.9	0.7	1.3	0.7	1	0.75
	10	2.1	1.8	1.4	2.6	1.4	2	1.5
	25	5.25	4.5	3.5	6.5	3.5	5	3.75
	50	10.5	9	7	13	7	10	7.5
	100	21	18	14	26	14	20	15
	150	31.5	27	21	39	21	30	22.5
	300	63	54	42	78	42	60	45
	600	126	108	84	156	84	120	90
	1200	252	216	168	312	168	240	180
	2000	420	360	280	520	280	400	300
	4000	840	720	560	1040	560	800	600
Cumulative dose	8440	1172.4	1519.2	1181.6	2194.4	1181.6	1688	1266

According to Niggeman's classification, the OFC clinical manifestations were classified as mild, moderate, and severe [21]. Moreover, for any reaction, we described the threshold dose of the culprit nut (the maximum tolerated dose during the OFC) and its corresponding dose of nut protein. The same classification was used to define the severity of the reaction in the clinical history.

Finally, the studied population was represented by children with confirmed nut allergies, comprising a clinical history of nut reactions, plus positive PbP and/or s-IgE to nut plus positive nut OFC. The diagnostic selection of children with nut allergies is shown in Figure 1. Subsequently, the patients were divided into groups according to the nut responsible for the first reaction in chronological order (according to clinical history). Then, demographic characteristics (gender, age, age at first nut reaction), coexisting allergic diseases (atopic dermatitis, asthma, rhinitis, another food allergy), familiar history of allergy, values of PbP to nut and s-IgE to nut, characteristics of OFC, and clinical manifestations of the first reaction to the nut were extrapolated through chart review. Finally, these characteristics were compared between the various nuts.

Statistical analyses were performed using OpenEpi (version 3.01; Atlanta, GA, USA), Microsoft Excel (2013 version, Redmond, WA, USA), and SPSS (IBM SPSS Statistics 22, Chicago, IL, USA). Qualitative data were expressed as counts and percentages; quantitative data were expressed as mean value $\pm$ standard deviation or median value and minimum–maximum value. Data distribution was verified with the Shapiro–Wilk test and equality of variance with the Hartley F-test. Differences between continuous variables were calculated using the Student's t-test and the one-way analysis of variance (ANOVA). Associations between categorical variables were obtained with Fisher's exact test and chi-square test. A *p*-value of less than 0.05 was considered statistically significant.

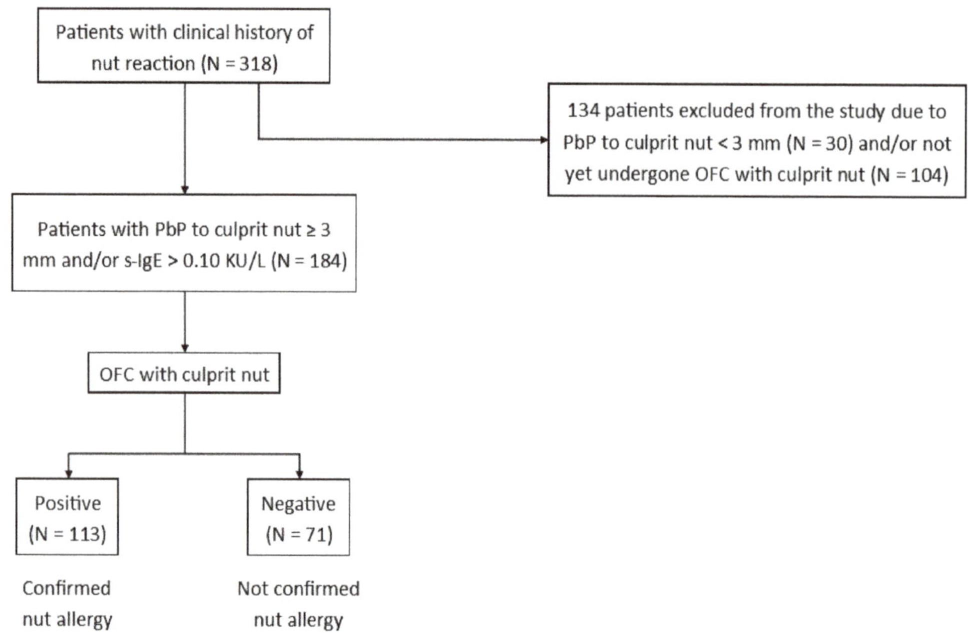

Figure 1. Flow chart used for the diagnostic selection of children with a confirmed nut allergy. Legend: N, number; OFC, oral food challenge; PbP, prick by prick; s-IgE, serum-specific IgE.

3. Results

We reviewed the clinical charts of 318 children (201 males (63%) and 117 females (37%)) with a clinical history of nut reaction to the following nuts according to the first reaction in chronological order: hazelnut (112; 35%), walnut (90; 28%), peanut (58; 18%), pine nut (30; 9%), cashew (14; 4%), pistachio (8; 3%), and almond (6; 2%). No patients reported reactions to macadamia, pecan, or Brazil nuts. Overall, 184 subjects had a positive PbP and/or s-IgE to nuts and underwent OFC with the culprit nut. The OFC was negative in 71 children (39%). Conversely, 113 subjects (61%) had a positive OFC and, therefore, a nut allergy was confirmed. The result of the OFC was independent of the type of nut tested ($p = 0.10$).

The demographic characteristics of 113 children with a confirmed nut allergy are reported in Table 2. According to the nut responsible for the first reaction considered in chronological order, we identified six different groups of patients: cashew (4; 4%), hazelnut (43; 38%), peanut (22; 19%), pine nut (11; 10%), pistachio (1; 1%), and walnut (32; 28%). The demographic characteristics of patients divided for the different nuts are summarized in Table 2.

There were no statistically significant differences in gender, concurrent allergic diseases, familiar history positive for allergic diseases, or mean values of PbP and s-IgE to nuts between the various nuts (Table 2). Conversely, the difference in mean age at the first nut reaction was statistically significant ($p = 0.00017$) for the pine nut group, which was higher (8.6 ± 3.7 years) than the mean age at the first reaction to the other nuts (hazelnut 3.7 ± 3.2 years, $p = 0.00008$; peanut 5 ± 3.4 years, $p = 0.009$; walnut 4.2 ± 2.7 years, $p = 0.0001$) except cashew ($p = 0.97$). Other differences in mean age were confirmed for the cashew (8.5 ± 5.9 years) and the hazelnut groups (3.7 ± 3.2 years; $p = 0.012$) but without evidence of statistical relevance with the remaining nuts.

Table 2. Demographic characteristics of patients with a nut allergy.

	Total (N = 113)	Cashew (N = 4)	Hazelnut (N = 43)	Peanut (N = 22)	Pine Nut (N = 11)	Pistachio (N = 1)	Walnut (N = 32)	p
Male (N = %)	74; 65	3; 75	33; 77	12; 55	5; 45	1; 100	20; 62	0.27
Age (months) (median; min; max)	42; 8; 175							
AD (N = %)	46; 41	0; 0	20; 47	6; 27	4; 36	1; 100	15; 47	0.22
Asthma (N = %)	33; 29	0; 0	14; 33	5; 23	5; 45	1; 100	8; 25	0.27
Rhinitis (N = %)	49; 43	1; 25	21; 49	7; 32	3; 27	1; 100	16; 50	0.38
Other FA (N = %)	34; 30	2; 50	17; 40	2; 9	2; 18	1; 100	10; 31	0.07
Family history of allergy (N = %)	85; 75	4; 100	33; 77	13; 59	9; 81	1; 100	25; 78	0.39
Age at first reaction (months) (mean ± SD; min; max)	57 ± 43; 8; 175	102 ± 71; 24; 172	45 ± 39; 8; 175	60 ± 41; 18; 154	103 ± 44; 48; 174	-	50 ± 33; 12; 125	0.00017
PbP (mm) (mean ± SD; min; max)	7 ± 3; 3; 15	8 ± 2; 6; 10	7 ± 3; 3; 15	6 ± 3; 3; 10	7 ± 2; 3; 10	-	7 ± 3; 3; 15	0.47
s-IgE (KU/L) (mean ± SD; min; max)	21 ± 32; 0.11; 100	3 ± 2; 1.7; 4.93	26 ± 35; 0.16; 100	31 ± 41; 0.12; 100	9 ± 19; 0.11; 66.3	-	14 ± 25; 0.3; 96.2	0.94

Legend: AD, atopic dermatitis; FA, food allergy; Max, maximum; Min, minimum; N, number; PbP, prick by prick; s-IgE, serum-specific IgE; SD, standard deviation; %, percentage.

Seventy-nine subjects (70%) denied other allergic food reactions, concurrent or previous, based on their clinical history. Among the 34 children with food co-allergies (30%), the most frequent was egg (17; 50%), followed by milk (14; 41%), fresh fruit (13; 38%), both egg and milk (11; 32%), fish/clams (9; 26%), cereals (3; 9%), legumes (2; 6%), and seeds (1; 3%). The frequency of food co-allergies did not differ between the six nuts groups (Table 2). As regards nut co-allergies, over 75% of subjects in every group, except for the pistachio one, reported a single nut reaction (Table 3), without statistical difference between groups (p = 0.45).

Table 3. Nut co-allergy.

	Other Nuts Allergies					
Nut	Cashew (N = %)	Hazelnut (N = %)	Peanut (N = %)	Pine Nut (N = %)	Pistachio (N = %)	Walnut (N = %)
Cashew (N = 4)	-	0; 0	0; 0	0; 0	1; 25	0; 0
Hazelnut (N = 43)	2; 5		3; 7	2; 5	1; 2	3; 7
Peanut (N = 22)	0; 0	2; 9		0; 0	0; 0	1; 5
Pine nut (N = 11)	1; 9	1; 9	0; 0		0; 0	2; 18
Pistachio (N = 1)	0; 0	1; 100	0; 0	0; 0		0; 0
Walnut (N = 32)	0; 0	6; 19	1; 3	0; 0	0; 0	

Legend: N, number; %, percentage.

Fifty-nine patients (52%) referenced a history of respiratory allergy (asthma and/or oculorhinitis). Grass pollen allergy was the most frequent among the pollen species tested (40; 68%), followed by cypress and birch (23; 59%).

Furthermore, 54 patients out of 113 (48%) had a history of anaphylaxis to nuts as the first reaction in chronological order: 25 patients (46% of anaphylaxis) reported moderate reactions while 19 (35% of anaphylaxis) reported severe ones. In 10 patients (19% of anaphylaxis), the severity of anaphylaxis at the first nut reaction was unknown. The occurrence of anaphylaxis (p = 0.16) and its severity at the first reaction in chronological order (moderate $p = 0.77$; severe $p = 0.10$) were not statistically different between the various nuts. The PbP and s-IgE values did not differ between the various nuts according to the severity of the first nut reaction. Moreover, in case of positive OFC, they did not differ among nuts according to the severity of the reaction (Table 4).

Table 4. Prick by prick and serum-specific IgE levels to the respective nuts, severity of the first nut reaction, and severity of the positive oral food challenge.

	Severity	Cashew (N = 4)		Hazelnut (N = 43)		Peanut (N = 22)		Pine Nut (N = 11)		Walnut (N = 32)		*p*	
		PbP (mm) (Mean ± SD)	s-IgE (KU/L) (Mean ± SD)	PbP (mm) (Mean ± SD)	s-IgE (KU/L) (Mean ± SD)	PbP (mm) (Mean ± SD)	s-IgE (KU/L) (Mean ± SD)	PbP (mm) (Mean ± SD)	s-IgE (KU/L) (Mean ± SD)	PbP (mm) (Mean ± SD)	s-IgE (KU/L) (Mean ± SD)	PbP	s-IgE
Oral food challenge / **First reaction**	Mild	-	-	7 ± 3	26 ± 35	7 ± 3	34 ± 45	7 ± 3	1 ± 1	6 ± 3	15 ± 28	1.31	0.56
	Moderate	8 ± 3	2 ± 1	9 ± 2	45 ± 47	5 ± 2	58 ± 60	8 ± 1	6 ± 5	-	-	0.20	0.27
	Severe	-	-	6 ± 3	13 ± 21	4 ± 2	26 ± 42	-	-	7 ± 4	2 ± 2	0.31	0.80
	Mild	8 ± 2	3 ± 2	7 ± 3	24 ± 35	6 ± 2	29 ± 41	7 ± 2	11 ± 23	6 ± 3	8 ± 11	0.52	0.78
	Moderate	-	-	6 ± 1	45 ± 34	6 ± 4	56 ± 43	-	-	9 ± 5	54 ± 49	0.40	0.84
	Severe	-	-	6 ± 3	-	-	-	6 ± 1	3 ± 3	6 ± 3	3 ± 3	0.97	0.76

Legend: N, number; PbP, prick by prick; s-IgE, serum-specific IgE; SD, standard deviation.

A complete molecular analysis was performed in 62 out of 97 eligible patients (64%): The highest adherence was obtained in the peanut group (86%), followed by the hazelnut (63%) and then walnut (50%) groups. Thus, the available molecular components were detected at least in 50% of the eligible population (Table 5). The mean values of molecular allergens did not correlate with the severity of the first nut reaction. Moreover, in case of positive OFC, they did not differ according to the severity of the reaction (Table 5).

Table 5. Molecular allergens, severity of the first nut reaction, and severity of the positive oral food challenge.

Molecular Allergens	Available Data (N=%)	Value (KU/L) (Mean ± SD; Min; Max)	First Reaction Mild (KU/L) (Mean ± SD)	Moderate (KU/L) (Mean ± SD)	Severe (KU/L) (Mean ± SD)	*p*	Oral Food Challenge Mild (KU/L) (Mean ± SD)	Moderate (KU/L) (Mean ± SD)	Severe (KU/L) (Mean ± SD)	*p*
Ara h 1	22; 100	33 ± 41; 0.15; 100	46 ± 51	28 ± 32	39 ± 40	0.80	34 ± 42	27 ± 45	-	0.98
Ara h 2	22; 100	38 ± 39; 0.6; 100	41 ± 47	32 ± 33	48 ± 42	0.92	34 ± 39	59 ± 45	-	0.64
Ara h 3	21; 95	19 ± 32; 0.11; 100	39 ± 43	11 ± 14	3 ± 3	0.88	20 ± 33	-	-	-
Ara h 8	21; 95	1 ± 1; 0.15; 2.17	1 ± 1	-	1 ± 1	0.53	0 ± 0	-	-	-
Ara h 9	21; 95	16 ± 22; 0.88; 40.8	16 ± 22	-	-	-	3 ± 4	-	-	-
Jug r 1	16; 50	11 ± 22; 0.27; 88.3	4 ± 4	-	-	-	13 ± 25	4 ± 5	4 ± 4	0.93
Jur r 3	16; 50	3 ± 6; 0.12; 16	4 ± 7	-	-	-	3 ± 7	-	-	-
Cor a 1	32; 74	8 ± 11; 0.16; 32.3	8 ± 13	12 ± 14	5 ± 7	0.42	7 ± 10	-	-	-
Cor a 8	32; 74	5 ± 11; 0.11; 36.6	3 ± 4	0 ± 0	20 ± 23	0.51	6 ± 11	-	-	-
Cor a 9	33; 77	22 ± 36; 0.11; 100	21 ± 33	50 ± 57	9 ± 18	0.14	20 ± 34	49 ± 20	-	0.92
Cor a 14	31; 71	16 ± 23; 0.11; 90	19 ± 27	16 ± 17	7 ± 10	0.08	17 ± 24	12 ± 12	-	0.95

Legend: Max, maximum; Min, minimum; N, number; SD, standard deviation; %, percentage.

During the first nut reaction, cutaneous involvement was the most frequent reaction (70; 62%), followed by gastrointestinal (44; 39%) and then respiratory clinical manifestations (25; 22%). Seven children (6%) reported signs and symptoms by contact. No one presented neurological involvement. Instead, five children (4%) referred to cardiovascular manifestations, described only in patients with anaphylaxis. During the OFCs, the signs and symptoms involved mainly the gastrointestinal system (82; 73%), followed by the cutaneous (42; 37%) and respiratory systems (28; 25%). No one presented cardiovascular or neurological involvement. The system involvement during the first nut reaction and the OFC is reported in Figure 2. The clinical features of the reactions did not differ between the various nuts (Figure 2), except for cutaneous involvement at the first nut reaction ($p = 0.011$), referred to mainly in the hazelnut and walnut groups.

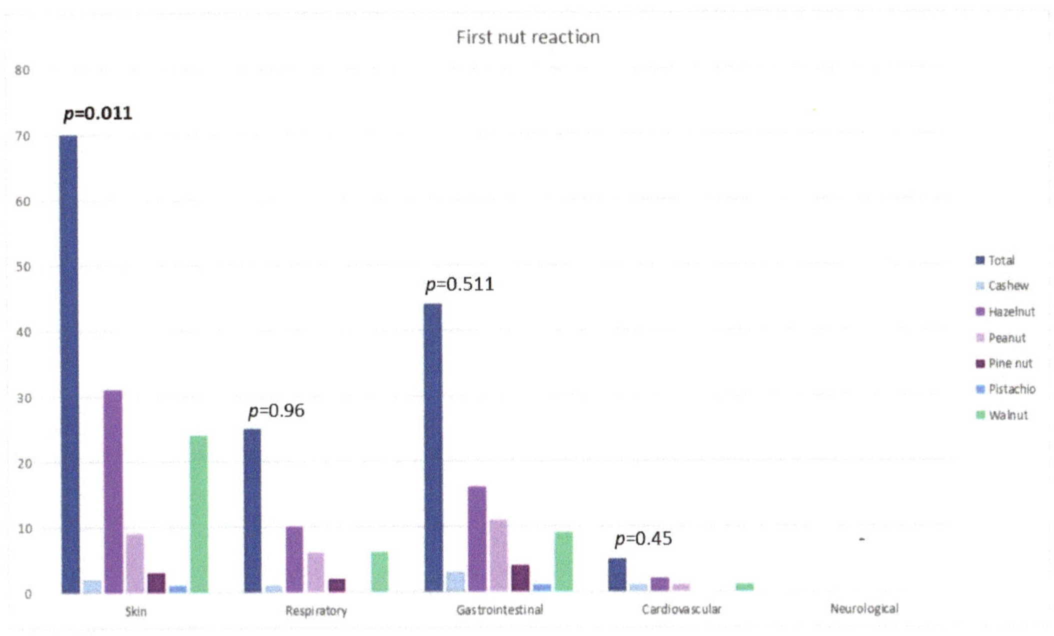

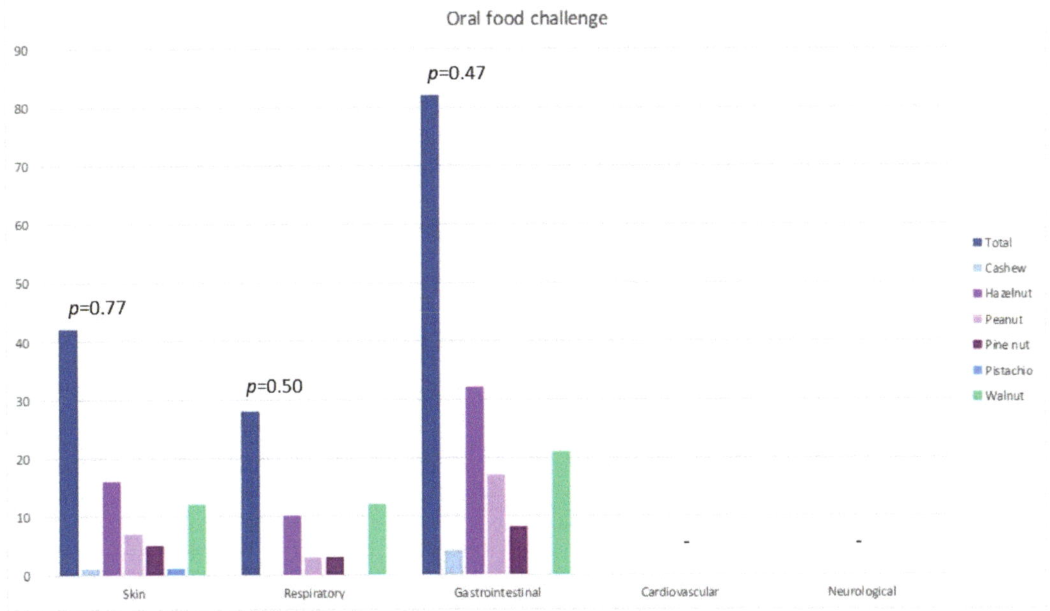

Figure 2. System involvement during the first nut reaction and the positive oral food challenge.

In total, 91 out of 113 subjects (81%) with clinical manifestations during OFC presented a mild reaction with single-system involvement. The remaining 22 subjects had anaphylaxis: moderate in 13 children (11%) and severe in 9 children (8%). Table 6 shows the severity of the reaction during the OFC. The severity of the reaction did not depend on the type of nut tested (Table 6).

Table 6. Severity of reaction of the positive oral food challenge and dose of nut ingested.

	Total (N = 113)	Cashew (N = 4)	Hazelnut (N = 43)	Peanut (N = 22)	Pine Nut (N = 11)	Pistachio (N = 1)	Walnut (N = 32)	p
Mild (N = %)	91; 81	4; 100	36; 84	18; 82	8; 73	1; 100	24; 75	0.77
Protein ingested (mean (mg) ± SD)	283 ± 630	79 ± 35	279 ± 518	472 ± 1054	199 ± 350	-	221 ± 478	0.65
Moderate (N = %)	13; 11	0; 0	5; 12	3; 14	1; 9	0; 0	4; 12	0.97
Protein ingested (mean (mg) ± SD)	376 ± 686	-	764 ± 1074	123 ± 75	-	-	168 ± 47	0.36
Severe (N = %)	9; 8	0; 0	2; 5	1; 5	2; 18	0; 0	4; 12	0.58
Protein ingested (mean (mg) ± SD)	69 ± 72	-	43 ± 32	-	104 ± 136	-	44 ± 56	0.64

Legend: N, number; %, percentage.

The mean dose of nut proteins ingested was 283 mg ± 630 mg (range 2–4500 mg) in mild reactions during OFC, 376 mg ± 686 mg (range 20–2500 mg) in moderate reactions, and 69 mg ± 70 mg (range 5–200 mg) in severe ones (Table 6). The dose of nut proteins ingested was significantly lower in severe reactions if compared with the mild ones ($p = 0.003$) but not with the moderate ones ($p = 0.13$). Furthermore, there were no statistically significant differences between the severity of reaction during the OFC and the mean dose of proteins ingested between the different nuts (mild, $p = 0.65$; moderate, $p = 0.36$; mild, $p = 0.64$) (Table 6).

Furthermore, 88 out of 113 children (78%) with a positive OFC showed a spontaneous resolution of signs and symptoms. The remaining 25 subjects (22%) with reactions during the OFC required the administration of therapy: 19 (76%) were treated with oral antihistamines, 20 (80%) with oral corticosteroids, and 3 (12%) with inhaled bronchodilators. None needed the administration of injectable epinephrine. None required hospitalization or intensive care assistance. The need for therapy did not depend on the type of nut tested ($p = 0.53$).

4. Discussion

To our knowledge, this is the first Italian study to provide a comprehensive characterization of children with nut allergies [15], although a Turkish study has recently been published on the same topic [22]. Indeed, most of the studies about nut allergies have focused on a single type of nut, especially peanut, walnut, or hazelnut, or summarized the main characteristics of nut allergies without making a distinction between the various kinds of nuts [11,23–26]. Conversely, we retrospectively analyzed the demographic, clinical, and allergological characteristics of Italian children with nut allergies and compared them.

From our experience, in Italian children, nut allergies are more common in male subjects, and allergies to hazelnuts and walnuts were the most observed nut allergies, as stated by a previous Spanish study [27]; peanut allergies were frequently observed as well. Furthermore, most of the first nut reactions occurred between 2 and 5 years (mean age 4.7 ± 3.6 years), which is later in comparison with those in the available literature [13,22]. The underlying reason for a later age of onset of the signs and symptoms may be due to the high percentage of familial history of allergy in our population, which leads the parents to introduce the nuts in the diet later for the fear of possible allergic reaction. The mean age of the first nut reaction was statistically higher for pine nuts (8.6 ± 3.7 years) when compared with the other nuts except for cashews (8.5 ± 5.9 years). On the contrary, in the hazelnut group, the mean age of the first nut reaction (3.7 ± 3.2 years) was lower than with other nuts, even without a statistical significance. These differences may be related to the low number of patients included in the study. The majority of patients in our population had a familiar history of allergy (75%) and/or concomitant allergic disorders (68%), including asthma, atopic dermatitis, and allergic rhinitis. Among these, allergic rhinitis was the most common allergic disorder (43%), followed by atopic dermatitis (41%). The percentage of children with allergic rhinitis and a concomitant food allergy is in line with the literature (33–40%) [28].

In subjects with food co-allergies, the most frequent foods involved were egg (50%), milk (41%), and concomitant egg and milk allergies (32%), according to the literature [22]. Most of our patients (over 75% in each group) reported a single nut allergy, and the remaining subjects had at least one nut co-allergy. A coexistent nut allergy was also described in several studies [29]. Sicherer et al. reported that 34% of patients allergic to peanuts or nuts might present with multiple nut allergies [30]; however, further studies reported a large variation in the proportion of patients reacting to multiple nuts, ranging from 12% to 96.7% [31]. We found more single nut allergies because our population is younger (median age of 3.5 years) than the other ones. In particular, the studies of Sicherer et al., Maloney et al., Mc William et al., and Brough et al. found values of co-allergies and median age, respectively, as follows: 34% and 3.6 years, 34% and 6.1 years, 47.8% and 6 years, and 60.7% and 5.5 years [29,30,32,33]. The only study that showed a lower percentage of co-allergy (12%) has a younger population (1.3 years) [34]. It seems that the percentage of the co-allergy increases with the increasing age of the studied population. In our study, the most common co-allergy was hazelnut, mainly represented in the group of patients allergic to walnut. These data could depend on the high prevalence of hazelnut allergy in continental Europe, in which it represents the most frequent nut allergy [8].

Anaphylaxis as the first clinical manifestation of nut allergy occurred in over 40% of our population, similarly to the Turkish study [22]. In agreement with the literature, the most common presenting signs and symptoms at initial diagnosis of nut allergy were skin manifestations (62%), including hives, itching, flushing, and/or rash [35,36]. The cutaneous involvement was followed by gastrointestinal (39%) and respiratory (2%) ones. In the same way, the characteristic of the population (gender, familiar history of allergic diseases, concurrent allergic diseases, food and/or nut co-allergy) did not differ between the various nuts, except for the mean age of the first nut reaction, as previously mentioned. Finally, we were unable to find a statistical association between the severity of nut reaction (first one in chronological order according to clinical history and during OFC), the mean values of allergological tests (PbP, s-IgE, molecular components), and the type of nut involved.

Among positive nut OFCs, we observed 22 severe reactions (19%). However, none of these required the injection of epinephrine, hospitalization, or intensive care assistance and the clinical manifestations resolved with oral antihistamines and corticosteroids. In the literature, the occurrence of anaphylaxis during the OFC depends on the type of food tested and, for TN and peanut, it ranges from 8% to 70% according to the different studies [37–39]. In the remaining population (81%), clinical manifestations during the OFC involved a single system (mainly the gastrointestinal one), according to different experiences carried out on a wide range of foods [40], and resolved spontaneously, confirming the safety of the OFC in children with suspected food allergies [36,38,39]. Thus, the clinical features, the severity of the reactions during the OFC, and the need for therapy did not depend on the type of nuts tested. Conversely, the mean dose of nut proteins ingested differed according to the severity of reactions during the OFC, with a lower threshold of doses observed in severe reactions.

The limitation of this observational study is the heterogeneous number of children retrospectively recruited for each kind of nut. These data could hide real differences between the various nut groups, but they could be informative as well, strictly connected to the characteristics of children with a nut allergy referred to our center, a tertiary-care pediatric hospital. However, as this study was carried out in a single allergy unit, the reference center for the region, our results could have limited applicability to other centers and regions. Another limitation is the lack of a complete molecular analysis of patients before the OFC was performed per clinical criteria. Therefore, only 64% of the eligible population received a molecular analysis; for this reason, it was not possible to clearly discriminate patients with primary nut allergy versus PFS/OAS. Finally, due to the small number of patients in the cashew and pistachio groups, the results concerning them should be interpreted with caution.

On the other hand, the strength of this study is the clear-cut selection of the studied population with an OFC-confirmed nut allergy and the summary of the main characteristics of the nut allergy, taking into account differences between the various kind of nuts. Finally, our study also confirmed the safety of an OFC performed by experienced personnel on selected subjects.

5. Conclusions

In conclusion, from our experience, the majority of children with an OFC-confirmed nut allergy have a familial history of allergy (75%) and/or concomitant allergic disorders (68%). Moreover, anaphylaxis is the first manifestation of nut allergy in a high percentage of children (48%), and the presence of anaphylaxis or severe reactions as the first clinical manifestation of a nut allergy does not differ among the various nuts. The clinical and allergological characteristics of children with nut allergies described in our study are similar to other international studies: hazelnut is the most frequent nut for referral. Over three out of four subjects have a single nut allergy. Finally, during the OFCs, the signs and symptoms involved mainly the gastrointestinal system (73%), with cases resolved spontaneously in most cases and infrequent severe reactions (19%), confirming the safety of OFCs in children with suspected food allergies. These results appear important to define a comprehensive characterization of children with nut allergies in Italy. However, the results should be confirmed by extensive data from international cohorts.

Author Contributions: S.T., S.B., T.A. and E.P. collected the data. S.T., S.B., M.G., G.L., L.S., G.P., C.A., E.N. and F.M. performed the investigations. S.T. and S.B. analyzed the data. S.T. and S.B. drafted the initial manuscript. S.T., S.B., M.G., G.L., L.S., G.P., C.A., E.N. and F.M. interpreted the data and reviewed the manuscript. All authors have read and agreed to the published version of the manuscript.

Funding: This research received no external funding.

Institutional Review Board Statement: The code of the event report issued by Meyer Children's University Hospital is: IR904-18-26854.

Informed Consent Statement: Written informed consent was obtained from the children's parents for all procedures performed.

Data Availability Statement: Aggregate analyses are available on reasonable request to the corresponding author.

Conflicts of Interest: The authors declare no conflict of interest.

Abbreviations

OFC	oral food challenge
PbP	prick by prick
s-IgE	serum-specific IgE
SPT	skin prick test
TN	tree nuts

References

1. Gupta, R.; Warren, C.; Smith, B.; Blumenstock, J.; Jiang, J.; Davis, M.; Nadeau, K. The public health impact of parent-reported childhood food allergies in the United States. *Pediatrics* **2018**, *142*, e20181235. [CrossRef]
2. Sicherer, S.H.; Munoz-Furlong, A.; Godbold, J.H.; Sampson, H.A. US prevalence of self-reported peanut, tree nut, and sesame allergy: 11-year follow-up. *J. Allergy Clin. Immunol.* **2010**, *125*, 1322–1326. [CrossRef]

3. Dunngalvin, A.; Dubois, A.E.J.; Blok, B.M.J.F.; Hourihane, J.O.B. The Effects of Food Allergy on Quality of Life. *Chem. Immunol. Allergy* **2015**, *101*, 235–252.

4. Primeau, M.; Kagan, R.; Joseph, L.; Lim, H.; Dufresne, C.; Duffy, C.; Prhcal, D.; Clarke, A. The psychological burden of peanut allergy as perceived by adults with peanut allergy and the parents of peanut-allergic children. *Clin. Exp. Allergy* **2000**, *30*, 1135–1143. [CrossRef]

5. Logan, K.; Du Toit, G.; Giovannini, M.; Turcanu, V.; Lack, G. Pediatric Allergic Diseases, Food Allergy, and Oral Tolerance. *Annu. Rev. Cell Dev. Biol.* **2020**, *36*, 511–528. [CrossRef] [PubMed]

6. Cetinkaya, P.G.; Buyuktiryaki, B.; Soyer, O.; Sahiner, U.M.; Sekerel, B.E. Factors predicting anaphylaxis in children with tree nut allergies. *Allergy Asthma Proc.* **2019**, *40*, 180–186. [CrossRef]

7. Eigenmann, P.A.; Lack, G.; Mazon, A.; Nieto, A.; Haddad, D.; Borugh, H.A.; Caubet, J.C. Managing Nut Allergy: A Remaining Clinical Challenge. *J. Allergy Clin. Immunol. Pract.* **2017**, *5*, 296–300. [CrossRef] [PubMed]

8. McWilliam, V.; Koplin, J.; Lodge, C.; Tang, M.; Dharmage, S.; Allen, K. The Prevalence of Tree Nut Allergy: A Systematic Review. *Curr. Allergy Asthma Rep.* **2015**, *15*, 54. [CrossRef] [PubMed]

9. Bock, S.A.; Muñoz-furlong, A.; Sampson, H.A. Fatalities due to anaphylactic reactions to foods. *J. Allergy Clin. Immunol.* **2001**, *107*, 191–193. [CrossRef]

10. de Leon, M.P.; Rolland, J.M.; O'Hehir, R.E. The peanut allergy epidemic: Allergen molecular characterisation and prospects for specific therapy. *Expert Rev. Mol. Med.* **2007**, *9*, 1–18. [CrossRef]

11. Stiefel, G.; Anagnostou, K.; Boyle, R.J.; Brathwaite, N.; Ewan, P.; Fox, A.T.; Huber, P.; Luyt, D.; Till, S.J.; Venter, C.; et al. BSACI guideline for the diagnosis and management of peanut and tree nut allergy. *Clin. Exp. Allergy* **2017**, *47*, 719–739. [CrossRef]

12. Walsh, J.; O'Flynn, N. Diagnosis and assessment of food allergy in children and young people in primary care and community settings: NICE clinical guideline. *Br. J. Gen. Pract.* **2011**, *61*, 473–475. [CrossRef]

13. Weinberger, T.; Sicherer, S. Current perspectives on tree nut allergy: A review. *J. Asthma Allergy* **2018**, *11*, 41–51. [CrossRef] [PubMed]

14. Elizur, A.; Appel, M.Y.; Nachshon, L.; Levy, M.B.; Epstein-Rigbi, N.; Golobov, K.; Goldberg, M. NUT Co Reactivity—ACquiring Knowledge for Elimination Recommendations (NUT CRACKER) study. *Allergy* **2018**, *73*, 593–601. [CrossRef]

15. Giovannini, M.; Comberiati, P.; Piazza, M.; Chiesa, E.; Piacentini, G.L.; Boner, A.; Zanoni, G.; Peroni, D.G. Retrospective definition of reaction risk in Italian children with peanut, hazelnut and walnut allergy through component-resolved diagnosis. *Allergol. Immunopathol.* **2018**, *47*, 73–78. [CrossRef]

16. Ortolani, C.; Ispano, M.; Pastorello, E.A.; Ansaloni, R.; Magri, G. Comparison of results of skin prick tests (with fresh foods and commercial food extracts) and RAST in 100 patients with oral allergic syndrome. *J. Allergy Clin. Immunol.* **1989**, *83*, 683–690. [CrossRef]

17. Heinzerling, L.; Mari, A.; Bergmann, K.-C.; Bresciani, M.; Burbach, G.; Darsow, U.; Durham, S.; Fokkens, W.; Gjomarkaj, M.; Haahtela, T.; et al. The skin prick test—European standards. *Clin. Transl. Allergy* **2013**, *3*, 3. [CrossRef]

18. Muraro, A.; Werfel, T.; Hoffmann-Sommergruber, K.; Roberts, G.; Beyer, K.; Bindslev-Jensen, C.; Cardona, V.; Dubois, A.; Dutoit, G.; Eigenmann, P.; et al. EAACI Food Allergy and Anaphylaxis Guidelines: Diagnosis and management of food allergy. *Allergy* **2014**, *69*, 1008–1025. [CrossRef] [PubMed]

19. Muraro, A.; Roberts, G.; Worm, M.; Bilò, M.B.; Brockow, K.; Fernández Rivas, M.; Santos, A.F.; Zolkipli, Z.Q.; Bellou, A.; Beyer, K.; et al. Anaphylaxis: Guidelines from the European Academy of Allergy and Clinical Immunology. *Allergy* **2014**, *69*, 1026–1045. [CrossRef] [PubMed]

20. Barni, S.; Liccioli, G.; Sarti, L.; Giovannini, M.; Novembre, E.; Mori, F. Immunoglobulin E (IgE)-Mediated Food Allergy in Children: Epidemiology, Pathogenesis, Diagnosis, Prevention, and Management. *Medicina* **2020**, *56*, 111. [CrossRef] [PubMed]

21. Niggemann, B.; Beyer, K. Time for a new grading system for allergic reactions? *Allergy* **2016**, *71*, 135–136. [CrossRef]

22. Cetinkaya, P.G.; Buyuktiryaki, B.; Soyer, O.; Sahiner, U.M.; Sackesen, C.; Sekerel, B.E. Phenotypical characterization of tree nuts and peanut allergies in east Mediterranean children. *Allergol. Immunopathol.* **2020**, *48*, 316–322. [CrossRef] [PubMed]

23. Abrams, E.M.; Chan, E.S.; Sicherer, S. Peanut allergy: New advances and ongoing controversies. *Pediatrics* **2020**, *145*, e20192102. [CrossRef]

24. Lyons, S.A.; Datema, M.R.; Le, T.M.; Asero, R.; Barreales, L.; Belohlavkova, S.; de Blay, F.; Clausen, M.; Dubakiene, R.; Fernández-Perez, C.; et al. Walnut Allergy Across Europe: Distribution of Allergen Sensitization Patterns and Prediction of Severity. *J. Allergy Clin. Immunol. Pract.* **2021**, *9*, 225–235.e10. [CrossRef]

25. Calamelli, E.; Trozzo, A.; Di Blasi, E.; Serra, L.; Bottau, P. Hazelnut allergy. *Medicina* **2021**, *57*, 67. [CrossRef] [PubMed]

26. Smeekens, J.M.; Bagley, K.; Kulis, M. Tree nut allergies: Allergen homology, cross-reactivity, and implications for therapy. *Clin. Exp. Allergy* **2018**, *48*, 762–772. [CrossRef]

27. Haroun-Díaz, E.; Azofra, J.; González-Mancebo, E.; de las Heras, M.; Pastor-Vargas, C.; Esteban, V.; Villalba, M.; Díaz-Perales, A.; Cuesta-Herranz, J. Nut allergy in two different areas of Spain: Differences in clinical and molecular pattern. *Nutrients* **2017**, *9*, 909. [CrossRef]

28. Boyce, J.A.; Assa'ad, A.; Burks, W.A.; Jones, S.M.; Sampson, H.A.; Wood, R.A.; Plaut, M.; Cooper, S.F.; Fenton, M.J.; Arshad, H.S.; et al. Guidelines for the diagnosis and management of food allergy in the United States: Report of the NIAID-sponsored expert panel. *J. Allergy Clin. Immunol.* **2010**, *126*, 51–58. [CrossRef]

29. Brough, H.A.; Caubet, J.-C.; Mazon, A.; Haddad, D.; Bergmann, M.M.; Wassenberg, J.; Panetta, V.; Gourgey, R.; Radulovic, S.; Nieto, M.; et al. Defining challenge-proven coexistent nut and sesame seed allergy: A prospective multicenter European study. *J. Allergy Clin. Immunol.* **2020**, *145*, 1231–1239. [CrossRef]
30. Sicherer, S.H.; Burks, A.W.; Sampson, H.A. Clinical Features of Acute Allergic Reactions to Peanut and Tree Nuts in Children. *Pediatrics* **1998**, *102*, e6. [CrossRef] [PubMed]
31. Midun, E.; Radulovic, S.; Brough, H.; Caubet, J.C. Recent advances in the management of nut allergy. *World Allergy Organ. J.* **2021**, *14*, 100491. [CrossRef]
32. Maloney, J.M.; Rudengren, M.; Ahlstedt, S.; Bock, S.A.; Sampson, H.A. The use of serum-specific IgE measurements for the diagnosis of peanut, tree nut, and seed allergy. *J. Allergy Clin. Immunol.* **2008**, *122*, 145–151. [CrossRef] [PubMed]
33. McWilliam, V.; Peters, R.; Tang, M.L.K.; Dharmage, S.; Ponsonby, A.L.; Gurrin, L.; Perrett, K.; Koplin, J.; Allen, K.J.; Dwyer, T.; et al. Patterns of tree nut sensitization and allergy in the first 6 years of life in a population-based cohort. *J. Allergy Clin. Immunol.* **2019**, *143*, 644–650.e5. [CrossRef]
34. Fleischer, D.M.; Conover-Walker, M.K.; Matsui, E.C.; Wood, R.A. The natural history of tree nut allergy. *J. Allergy Clin. Immunol.* **2005**, *116*, 1087–1093. [CrossRef] [PubMed]
35. Couch, C.; Franxman, T.; Greenhawt, M. Characteristics of tree nut challenges in tree nut allergic and tree nut sensitized individuals. *Ann. Allergy, Asthma Immunol.* **2017**, *118*, 591–596. [CrossRef]
36. Anagnostou, K. Safety of Oral Food Challenges in Early Life. *Children* **2018**, *5*, 65. [CrossRef] [PubMed]
37. Abrams, E.M.; Becker, A.B. Oral food challenge outcomes in a pediatric tertiary care center. *Allergy Asthma Clin. Immunol.* **2017**, *13*, 43. [CrossRef]
38. Järvinen, K.M.; Amalanayagam, S.; Shreffler, W.G.; Noone, S.; Sicherere, S.H.; Sampson, H.A.; Nowak-Wegrzyn, A. Epinephrine treatment is infrequent and biphasic reactions are rare in food-induced reactions during oral food challenges in children. *J. Allergy Clin. Immunol.* **2009**, *124*, 1267–1272. [CrossRef]
39. Lieberman, J.A.; Cox, A.L.; Vitale, M.; Sampson, H.A. Outcomes of office-based, open food challenges in the management of food allergy. *J. Allergy Clin. Immunol.* **2011**, *128*, 1120–1122. [CrossRef]
40. Ballini, G.; Gavagni, C.; Guidotti, C.; Ciolini, G.; Liccioli, G.; Giovannini, M.; Sarti, L.; Ciofi, D.; Novembre, E.; Mori, F.; et al. Frequency of positive oral food challenges and their outcomes in the allergy unit of a tertiary-care pediatric hospital. *Allergol. Immunopathol.* **2021**, *49*, 120–130. [CrossRef]

Review

Phenotypes and Endotypes of Peach Allergy: What Is New?

Simona Barni [1], Davide Caimmi [2,3], Fernanda Chiera [4], Pasquale Comberiati [5,6,*], Carla Mastrorilli [7], Umberto Pelosi [8], Francesco Paravati [4], Gian Luigi Marseglia [9] and Stefania Arasi [10]

[1] Allergic Unit, Department of Pediatric, Meyer Children's Hospital, 50139 Florence, Italy; simonabarni@hotmail.com
[2] Allergy Unit, CHU de Montpellier, Université de Montpellier, 34295 Montpellier, France; davide.caimmi@gmail.com
[3] IDESP, UMR A11–INSERM, Université de Montpellier, 34093 Montpellier, France
[4] Department of Pediatrics, San Giovanni di Dio Hospital, 88900 Crotone, Italy; fernandachiera@hotmail.it (F.C.); paravati.f@gmail.com (F.P.)
[5] Department of Clinical and Experimental Medicine, Section of Pediatrics, University of Pisa, 56126 Pisa, Italy
[6] Department of Clinical Immunology and Allergology, I.M. Sechenov First Moscow State Medical University, 119991 Moscow, Russia
[7] Department of Pediatrics, University Hospital Consortium Corporation Polyclinic of Bari, Pediatric Hospital Giovanni XXIII, 70124 Bari, Italy; carla.mastrorilli@icloud.com
[8] Pediatric Unit, Santa Barbara Hospital, 09016 Iglesias, Italy; umberto.pelosi@gmail.com
[9] Department of Pediatrics, University of Pavia, San Matteo Foundation IRCCS Policlinico, 27100 Pavia, Italy; gl.marseglia@smatteo.pv.it
[10] Area of Translational Research in Pediatric Specialities, Allergy Unit, Bambino Gesù Children's Hospital, IRCCS, 00165 Rome, Italy; stefania.arasi@opbg.net
* Correspondence: pasquale.comberiati@gmail.com

Citation: Barni, S.; Caimmi, D.; Chiera, F.; Comberiati, P.; Mastrorilli, C.; Pelosi, U.; Paravati, F.; Marseglia, G.L.; Arasi, S. Phenotypes and Endotypes of Peach Allergy: What Is New? *Nutrients* 2022, 14, 998. https://doi.org/10.3390/nu14050998

Academic Editor: Margarida Castell

Received: 21 January 2022
Accepted: 25 February 2022
Published: 26 February 2022

Publisher's Note: MDPI stays neutral with regard to jurisdictional claims in published maps and institutional affiliations.

Abstract: Peach allergy is emerging as a common type of fresh-fruit allergy in Europe, especially in the Mediterranean area. The clinical manifestations of peach allergy tend to have a peculiar geographical distribution and can range from mild oral symptoms to anaphylaxis, depending on the allergic sensitization profile. The peach allergen Pru p 7, also known as peamaclein, has recently been identified as a marker of peach allergy severity and as being responsible for peculiar clinical features in areas with high exposure to cypress pollen. This review addresses the latest findings on molecular allergens for the diagnosis of peach allergy, the clinical phenotypes and endotypes of peach allergy in adults and children, and management strategies, including immunotherapy, for peach allergy.

Keywords: peach allergy; food allergy; molecular allergy; Pru p 3; Pru p 7; peamaclein; anaphylaxis; oral allergy syndrome; pollen-food allergy syndrome; oral immunotherapy

1. Epidemiology

Peach (*Prunus persica*) is the fruit of *Prunus* trees, belonging to the *Rosaceae* family, including 4828 known species in 104 genera [1–3]. In addition to peaches, apples, pears, quinces, apricots, plums, cherries, raspberries, loquats, strawberries, and almonds belong to the *Rosaceae* family [4].

Currently, the peach plant is cultivated in different parts of the world. Peach cultivation is believed to have originated in China and to have been transported, via the silk route, to India, the Middle East, and Persia, before finally spreading towards Europe. China, Italy, Spain, Turkey, and the USA are the leading peach-producing countries [5].

Peach fruit usually ripens between August and September [1]. Peach can be eaten as fresh fruit as well as in treated forms, such as canned, dried, juice, and jam [5].

Peach has been described as a common cause of fresh-fruit allergy in Europe, especially in the Mediterranean area [6]. The prevalence data on fruit allergies are limited, and the available data are derived from scarce studies, especially in children [7–9]. In a systematic review conducted by Zuidmeer et al., the overall perceived prevalence of fruit allergies ranged from 0.1% to 4.3% [7]. In particular, 2.2–11.5% of children aged 0–6 years and 0.4–6.6% of adults are affected by fruit allergies, based on self-reported data. One European-based large survey reported the highest and lowest prevalence of allergic sensitization to peach in Germany (11.7%) and Iceland (0.3%), respectively [8]. In another, similar survey, the highest prevalence of peach sensitization was observed in Switzerland (13.4%) and the lowest in Iceland (2.3%) [9]. The overall European prevalence of allergic sensitization to peach increased from 5.4% in 2010 [8] to 7.9% in 2014 [9]. The prevalence data on peach-allergen-specific sensitization have been investigated in Spanish and Italian studies: lipid-transfer protein (LTP) sensitization is predominant in Southern Europe, whereas sensitization to pathogenesis-related 10 (PR-10) is more common in Northern and Central Europe, including areas with Fagales pollen exposure (birch, alder, hazel, hornbeam, oak, beech, and chestnut) [10,11].

Similar to other IgE-mediated food allergies, peach allergy negatively impacts quality of life, causing stress and anxiety. Peach allergy, as with fruit allergies in general, is reported to be associated with less-severe symptoms than food allergies to peanuts and tree nuts; nevertheless, the condition exerts a similar impact on patients' quality of life: 60% of adults are impacted by fruit allergy in their daily life at home and 62% in their life outside the home [12].

2. Peach Allergens

To date, six peach allergens have been recognized [13]. Detailed information on each allergenic protein is provided in Table 1 [14–39].

Table 1. Main features of peach molecular allergens. Modified from [35].

Allergen	Biochemical Name	Molecular Weight (kDa)	Main Characteristic
Pru p 1	Pathogenesis-related protein group 10, (PR-10), Bet v 1 family member	18	Mainly found in areas with high birch pollen exposure [10].
Pru p 2	Thaumatin-like protein (TLP)	25–28	Pru p 2 from peach was one of the probable allergens causing fruit allergies [36].
Pru p 3	Non-specific lipid-transfer protein 1 (nsLTP1)	10	Major allergen [10]. Present in the outer surface of peach [27]. In total, 54 (96%) out of 57 children showed positive Pru p 3-sIgE in a Spanish study [10].
Pru p 4	Profilin	14	Minor allergen [10]. In total, 52 (12.1%) out of 430 patients were sensitized to profilins in an adult study [37].
Pru p 7	Gibberellin-regulated protein (GRP)	6910.84 Da (Mass spectrometry)	Major allergen [33]. Identified in 2012 [13]. Present both in the pulp and in the peel [27]. Sensitization to Pru p7 was present in 171 (54%) out of 316 subjects with suspected peach allergy [33]. Pru p 7 sensitization was more frequent in peach-allergic (123/198, 62%) than in peach-tolerant (48/118, 41%) patients, *p*-value = 0.0002 [33].
Pru p 9	Pathogenesis-related protein PR-1	18	Identified in 2018 [13]. Sensitization to peach-tree pollen was rated third, after olive tree and grass [38], in areas with peach-tree cultivars. In total, 205 (30%) out of 685 children were sensitized to Pru p 9 on skin prick test [38].

kDa: kilodaltons, IgE: immunoglobulin E; sIgE: specific IgE.

2.1. Pru p 1

Pru p 1 is a member of the PR-10 protein family and is present in the pulp and the skin of peach [14]. It shares a structural homology with the major birch pollen, Bet v1 [15]. For this reason, Pru p1 sensitization is commonly found in Northern and Central Europe, where the exposure to birch pollen is high and usually results in oral allergy syndrome (OAS) symptoms [15]. Pru p1 cross-reacts with other PR-10 protein families, such as *Rosaceae* fruits, hazelnut, carrots, and celery [16]. Pru p 1 is heat-labile and it is found to be sensitive to gastrointestinal digestion [14]. Thus, only the unprocessed form of the fruit leads to the typical symptoms of OAS, whereas cooked peach is tolerated by patients [16].

2.2. Pru p 3

Pru p 3 is a non-specific LTP (nsLTP) [17]. The outer surface of the peach (pericarp) has a high concentration of nsLTP [18]. The peel contains seven times higher LTP than the pulp [16]. Pru p 3 cross-reacts with the nsLTP contained in the other fruits of the *Rosaceae* family (apple, plum, cherry and apricot), as well as in vegetables (asparagus, lettuce, tomato, maize, onion, and carrot) and nuts (walnut, hazelnut, almond and peanut) [19]. The LTP nsLTP is a plant panallergen due to its widespread distribution among plant-foods and pollens [16]. The LTPs from different plant-food and pollens can cross-react with each other, causing sensitization and, eventually, symptoms in multiple plant foods, a condition also known as "LTP syndrome" [20]. Pru p 3 is resistant to heat and proteolytic digestion. Therefore, the clinical manifestations of Pru p 3 sensitization can range from mild OAS symptoms to severe systemic allergic reactions (anaphylaxis) [21,22].

2.3. Pru p 4

Pru p 4 is a profilin, which is an important pan-allergen, widely found in pollens and vegetables [23]. Pru p 4 is present in the pulp and peel of peach [10]. Pru p 4 cross-reacts with profilins from other members of the *Rosaceae* family (i.e., apple and cherry) and with profilin from unrelated families' pollen (i.e., *Artemisia vulgaris*, *Betula alba*, *Corylus avellanus*, *P amygdalus*) [24]. Pru p 4 is heat-labile and it can be destroyed by gastrointestinal digestion [25]. For this reason, the usual clinical manifestation of Pru p 4 sensitization is OAS [24,26].

2.4. Pru p 7

Pru p 7 is a gibberellin-regulated protein (GRP) [27], also known as Snakin/GASA [28], that was first described by Tuppo et al. in 2013 [27]. Pru p 7 has been found both in the pulp and the peel of the peach [27]. Pru p 7 cross-reacts with several fruits of the *Rosaceae* (i.e., apricot and pomegranate) and *Rutaceae* family (i.e., orange), as well as pollens from the *Cupressaceae* family [29–32]. Indeed, Pru p 7 sensitivity seems to be most common in areas with high cypress pollen exposure [33]. Pru p 7 is resistant to heat and proteolytic digestion [27]. Thus, the typical allergic symptom of Pru p 7 sensitization is anaphylaxis. Similar to Pru p 3, sensitization to Pru p 7 is considered a risk factor for severe allergic reactions to fresh fruit [33]. Biagioni et al. [34] recently reported the first case series of children with documented Pru p 7 allergies and provided a diagnostic algorithm. The authors suggest performing skin prick tests (SPT) for inhalant and food allergens, including both cypress pollen and Pru p 3-enriched peach peel extracts, in case of a systemic allergic reaction to fruit. In cases of a positive SPT for both cypress- and Pru p 3-enriched peach peel extract and a negative in vitro result for specific IgE (sIgE) to Pru p 3, the diagnosis of Pru p 7 allergy is highly probable. In these cases, whenever possible, determining serum sIgE levels of Pru p 7 is recommended.

2.5. Pru p 9

Pru p 9 is a pathogenesis-related protein PR-1, identified in 2018 [13], with a molecular weight of 18 kDa. In 685 Spanish children and adolescents affected by rhino-conjunctivitis and asthma, the sensitization to peach-tree pollen was rated third, after olive tree and grass. Thirty percent (205 out of 685) of children were sensitized to Pru p 9 on skin prick testing [38]. The rate of sensitization to Pru p 9 in children is similar to that in adults from the same area [39]. Pru p 9 is considered a new occupational allergen from peach-tree pollen involved in rhinitis and asthma [39].

3. Clinical Manifestations

Similar to other IgE-mediated food allergic reactions, symptoms appear within minutes to two hours from peach ingestion, except for food-dependent exercise-induced anaphylaxis, which can occur up to 4 h later. Reactions can be triggered by the allergen through the oral route, rarely by inhalation or skin contact, and may affect one or more target organs, including the oral mucosa, the skin, the gastrointestinal tract, the respiratory tract, and the cardiovascular system [40–42].

Immediate peach-induced reactions could be associated with two clinical patterns: the pollen-food allergy syndrome (PFAS) and a primary food allergy [33].

The clinical manifestations of peach allergy depend on the sensitization profile and, consequently, have a peculiar geographical distribution.

In Northern and Central Europe, peach allergy is mainly secondary to pollen allergy. In this condition, also known as PFAS, pollen allergens are the causative agents of the primary sensitization and food allergy to fruits and vegetables results from cross-reactivity between pollen and food allergens. Conversely, in Mediterranean countries, fruit allergy without related pollinosis is often observed and systemic reactions are frequently reported [43,44].

While, on one hand, it is true that allergy to Pru p 1 is mainly associated with pollen-fruit allergy syndrome, and that Pru p 9 allergy is associated with respiratory symptoms, on the other hand, patients allergic to either Pru p 3 and/or Pru p 7 are at risk of developing severe symptoms, including anaphylaxis and fatal anaphylaxis [33,45,46].

3.1. Peach Allergy Secondary to Pollen Allergy

The allergen families involved in peach-induced PFAS include PR10 proteins, profilins, nsLTPs, thaumatin-like proteins, and gibberellin-regulated proteins [47].

PFAS account for up to 60% of food allergies in adult patients and adolescents. It may affect one or more target organs: the skin, the oral mucosa, the gastrointestinal tract, the respiratory tract, and the cardiovascular system [47,48].

The most frequent clinical pattern observed in adult patients and adolescents with PFAS is OAS. Symptoms emerge within 5–15 min of food ingestion and consist of tingling/itching of the lips, tongue, oral mucosa, palate, and throat, with possible mild angioedema associated at the same sites [48].

Most cases resolve spontaneously within 30 min, but 3% of patients present systemic reactions without oropharyngeal symptoms, and 1–8% develop systemic reactions, such as urticaria, dyspnea, wheezing, and anaphylaxis [49–51].

Acute generalized urticaria, with or without angioedema, and contact urticaria are the second most frequently observed symptoms of PFAS. Gastrointestinal symptoms, such as nausea, vomiting, abdominal pain, and diarrhea are rarely seen as exclusive manifestations of PFAS. Respiratory symptoms, such as rhinoconjunctivitis, bronchospasm, and laryngeal edema occur more frequently in association with other target organs symptoms rather than in isolation [48].

The presence of comorbidities (atopic dermatitis) and cofactors (exercise, alcohol consumption, use of non-steroidal anti-inflammatory drugs (NSAIDs)) increases the severity of symptoms and the risk of anaphylaxis [50].

3.2. Primary Peach Allergy

Primary food allergy to peach, in which the sensitization occurs through the ingestion of the food, is mainly related to nsLTP Pru p 3, although some studies reported primarily airborne sensitization to nsLTPs [52,53].

In the Mediterranean area, there is a high rate of sensitization to nsLTPs, which represents the most frequent cause of both primary food allergy and food-dependent anaphylaxis in adults living in these countries [54,55].

The sensitization to Pru p 3 often occurs early in life. It may be isolated (mono-sensitization) or associated with multiple nsLTP sensitizations, which may lead to multiple plant-food allergies (nsLTP-syndrome) [56].

Pru p 3 sensitization may be asymptomatic or manifest with variable symptom severity, ranging from OAS to anaphylaxis [57,58].

OAS and contact urticaria are the most frequent clinical patterns observed in LTP hypersensitivity. Gastrointestinal symptoms (nausea, vomiting, abdominal pain, diarrhea) may occur as isolated symptoms or in association with the cutaneous, respiratory, or cardiovascular symptoms involved in anaphylaxis [57].

A study on LTP syndrome reported that in a group of 87 patients sensitized to Pru p 3, 44% had anaphylaxis, 43% presented OAS or urticaria, and 13% were asymptomatic. The culprit food belonged to the *Rosaceae* family in 48.8% of the subjects, and the most frequent food involved was peach in both symptomatic groups [59].

Co-sensitization to birch pollen (Bet v 1) and/or to profilin is associated with a lower prevalence of severe reactions and a higher prevalence of local reactions (OAS) [58].

A large prospective study evaluated the phenotype and severity biomarkers of peach-allergic patients sensitized to Pru p 3. The authors showed that most patients were sensitized to other LTP-containing plant foods (LTP syndrome), while only 6.8% were LTP-monoallergic (reacting only to peach and not to other plant foods). Subjects with LTP syndrome had a younger onset of peach allergy, and more asthma and sensitization to Parietaria and profilin than the LTP-monoallergic patients. Anaphylaxis was significantly more frequent in the LTP-monoallergic group, which had no sensitization to profillin. The presence of profilin sensitization was associated with a lower risk of anaphylaxis. No correlation was observed between SPT diameter, Pru p 3 sIgE level, level of nsLTP sensitization, and severity of reaction to peach [60].

Individuals with sensitization to Pru p 3 may develop cross-sensitization to other nsLTPs containing plant foods due to the structural homology between different nsLTPs. Pru p 3 shows a sequence homology from 62% to 81% with analog proteins from apple (Mal d 3), apricot (Pru ar 3), plum (Pru d 3), cherry (Pru av 3), orange (Cit s 3), strawberry (Fra a 3), and grape (Vit v 1). Other LTPs with a structural homology with Pru p 3 are present in peanut (Ara h 9), wheat (Tri a 14), hazelnut (60% with Cor a 8), and walnut (66% with Jug r 3) [61,62]. The risk of cross-reactivity most frequently involves the fruits of the Rosaceae family (apple, plum, apricot, cherry), but also nuts and peanuts. The clinical pattern ranges from local oropharyngeal symptoms up to anaphylaxis [62].

Co-factors are often involved (up to 40% of cases) in clinical expression: fasting, exercise, menstruation, and NSAID could determine the appearance of symptoms in patients sensitized to nsLTPs or influence symptom severity. According to Pascal et al., a cofactor is identified as precipitating anaphylaxis in 32.4% of nsLTPs allergic patients [20].

Sensitization nsLTPs could also be involved in food-dependent exercise-induced anaphylaxis (FDEIA), provoked by the combination of food ingestion and physical exercise within 4 h of food ingestion and within one hour of the start of exercise [63].

In patients with peach-FDEIA, Pru p 3 is the most frequent sensitizer, followed by Pru p 7 [56,63,64].

3.3. Peamaclein Allergy

The peach allergen Pru p 7, also known as peamaclein, has recently been identified as a marker of peach allergy severity and as being responsible for peculiar clinical features, sometimes occurring in the presence of cofactors [33,65].

Peamaclein allergy is mostly observed in adolescents and adults. Pru p 7, similarly to Pru p 3, resists heat and digestion and it is suspected to cause a primary food allergy through the gastrointestinal tract route [29]. However, a recent study reported that sensitization to Pru p 7 develops in areas with high exposure to cypress pollen, due to the homology between Cypmaclein and Pru p 7, inducing a PFAS syndrome more severe than those previously described [33,65].

Moreover, Pru p 7 presents homology with Pru m 7 (Japanese apricot), Pun g 7 (pomegranate), Pru av 7 (cherry), and Cit s 7 (orange). In particular, Pru p 7 shows 100% sequence homology with Pru m 7, 97% with Pru av 7, 90% with Pun g 7, 87% with Cit s 7, 84% with black cottonwood GRP, 82% with potato GRP, and 81% with soybean GRP [66]. The clinical cross-reactivity between GRPs was reported among peach, Japanese apricot, orange, and pomegranate. In addition to these fruits, patients with GRP sensitization frequently experience allergic reactions against apple due to the presence of a GRP named applemeclein. It shares a 94% homology with Pru p 7 (peamaclein), Pru m 7 (Japanese apricot), and Pru av 7 (cherry) [67].

A recent multicenter study, including 316 subjects from France, reported that sensitization to Pru p 7 is common in peach-allergic subjects, with a prevalence of 62%, and it occurs often as monosensitization (54%). Furthermore, Pru p 7 sensitization and sIgE levels were higher in patients experiencing Grade 3 reactions, according to EAACI classification [33,68].

Swelling of the face, especially the eyelids, oropharyngeal tightness, and anaphylaxis featured with peamaclein allergy [29].

Inomata et al. observed, among peach-allergic patients sensitized to Pru p 7, that the most frequent symptoms were oropharyngeal (69.2%), followed by laryngeal tightness (46.2%), facial edema (46.2%), eyelid edema (46.2%), urticaria (38.5%), dyspnea (23.1%), nasal obstruction (23.1%), conjunctival injection (15.4%), lip edema (15.4%), loss of consciousness (15.4%), and hypotension (7.7%) [69].

4. Diagnosis

As with any diagnostic workup for food allergy, screening allergen-sIgE testing without clinical necessity is discouraged [40,47,70]. A detailed clinical history is therefore crucial for selecting the appropriate confirmatory tests. According to the ICON and EAACI guidelines for food allergies [40], the diagnosis of peach allergy lies on the combination of a convincing clinical history of immediate reaction to peach and positive IgE sensitization testing assessed through SPT to peach (in the form of either extract, molecular components, or fresh peach), and/or IgE sensitization to peach (either extracts or molecular components). Where the diagnosis is unclear, an oral food challenge (OFC) is required as the gold standard test to provide a definitive diagnosis and to prevent patients from unneeded and potentially harmful elimination diets. However, OFC is logistically demanding, and anaphylactic reactions may occur. Reliable prognostic markers or algorithms integrating different clinical and biological parameters for predicting the severity of allergic reactions during OFC are under investigation.

4.1. Clinical History

A convincing clinical history is usually defined as one or more immediate reaction(s) within two hours of peach ingestion, inhalation, or direct contact, presenting as acute urticaria or angioedema, contact urticaria, laryngeal swelling, immediate vomiting, rhinitis, cough, wheezing, bronchospasm, hypotension or loss of consciousness, oral allergy syndrome (i.e., itching and tingling of the lips, oral mucosa and/or tongue), or food-dependent exercise-induced anaphylaxis [40,48,70]. The severity of reactions is useful for suspecting specific patterns of sensitization and proper management. Peamaclein (Pru p 7) frequently elicits anaphylaxis [71] and, similarly to allergy to other gibberellins, often includes peculiar clinical symptoms, such as facial swelling and laryngeal tightness, which can be predictive factors for gibberellin allergies [29]. Because of their labile chemical structure, profilins (Pru p 4 in peach) and PR-10 (Pru p 1 in peach) are usually responsible for mild symptoms [72]. Cofactors should be always investigated (e.g., asthma exacerbations, infections, exercise, alcohol, tiredness, use of NSAIDs, and menstruation), since they usually play a crucial role in eliciting reactions in patients allergic to nsLTP (Pru p 3 in peach) and, less frequently, in patients allergic to gibberellins (more evident for Pru m 7 (apricot) and Cit s 7 (orange) less for Pru p 7 (peach)). In t patients who have peach-FDEIA, nsLTPs are the most frequent sensitizers, followed by peamaclein [56].

Therefore, the clinical history should include the following: possible causative food(s) (peach and other fruits/vegetables), the time of onset, the extent and reproducibility of symptoms, the identification of allergic symptoms with plants and plants food(s), the quantity of ingested food, details of the food preparation (e.g., raw vs. cooked, peeled vs. unpeeled), and the relevance of cofactors.

4.2. IgE Sensitization

The use of peach-specific IgE determination combined with clinical history and peach SPT may reduce the need for OFC [73]. Component-resolved diagnosis (CRD), which uses single allergenic components for the assessment of epitope-sIgE, can provide critical information for predicting individualized sensitization patterns and the risk of severe allergic reactions [72]. Only molecular diagnostics makes it possible to identify and differentiate sensitization to LTP or peamaclein. Peach LTP extracts for SPT are contaminated with peamaclein Pru p 7, because LTP Pru p 3 and peamaclein Pru p 7 have similar molecular weights.

The use of commercial peach extracts for SPT is useful in clinical practice. However, clinicians should consider that peach extract for SPT most likely lacks labile peach allergens (i.e., Pru p 1, and Pru p 4), because these are usually lost during production procedures. By contrast, stable allergens, such as Pru p 3 and Pru p 7, are usually retained in commercial peach extracts [27]. Consequently, SPT with current extracts may furnish a prompt, first-level, component-resolved diagnosis at the bedside [74,75].

The use of serum-sIgE against molecular components provides useful support to the diagnosis and may help with risk stratification, assessment, and management. Pru p 7 is a small protein that is upregulated upon biotic stress. It represents a major allergen associated with severe clinical symptoms and strong cypress pollen sensitization [33]. A study conducted in the southern part of France evaluated 316 patients with suspected peach allergy. According to the ICON and EAACI guidelines for food allergies, peach allergy was diagnosed in 198 subjects. Sensitization to Pru p 7 was present in 171 (54%) of all the subjects in the study and 123 of 198 (62%) were diagnosed as peach-allergic, more than half of whom were sensitized to no other peach allergen. The frequency and magnitude of Pru p 7 sensitization were associated with the presence of a peach allergy, the clinical severity of peach-induced allergic reactions, and the level of cypress pollen exposure. Cypress pollen extract completely outcompeted IgE binding to Pru p 7 [35].

4.3. Oral Food Challenge

If the diagnosis of peach allergy is in doubt, OFC is required as it represents the gold standard for the diagnosis of any food allergy. Some OFC protocols are intended to test peach peel and pulp separately [33], others to test them both, and some to assess exercise-induced anaphylaxis [76]. Furthermore, clinicians may consider allergy testing and, ultimately, OFC to plant foods containing nsLTPs or GRP with known potential cross-reactivity with peach if oral tolerance to these foods is in doubt, and according to the patient's preference (Figure 1).

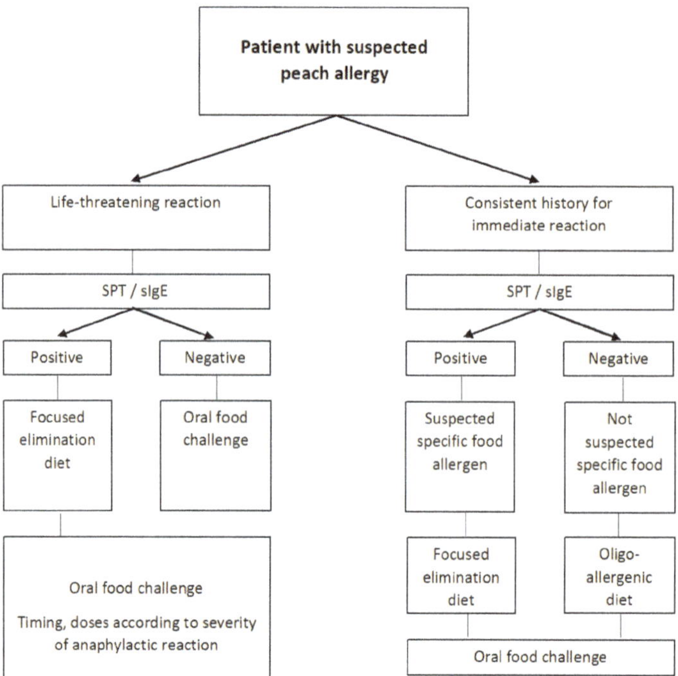

Figure 1. Diagnostic algorithm for peach allergy. Adapted from [40].

5. Prevention and Management

5.1. Primary and Secondary Prevention

To date, no study has shown a possible effective strategy for the primary prevention of peach allergy. Neither polyunsaturated fatty acid supplementation during pregnancy nor the use of probiotics and fish oil supplementation in infancy were effective at preventing the appearance of food allergies [77]. The early introduction of food during diversification could be a possible primary prevention strategy [78]. Even though current data show moderate evidence that the early introduction of peanut and egg reduces the risk of food allergy, there is no sufficient scientific information on other major food allergens [79].

In peach-allergic patients, as with other food allergies, prescribing preventive antihistamines was not shown to be effective at preventing possible allergic reactions; furthermore, this strategy may delay the timely and appropriate use of adrenaline to treat anaphylaxis [40]. The use of mast-cell stabilizers to prevent allergic reactions showed different clinical results, making it not advisable, so far, as a prophylactic strategy for food allergies in general and, therefore, for peach allergy as well [40]. The use of monoclonal antibodies, such as omalizumab and dupilumab, has been suggested instead in the treatment of food allergies, mostly as adjuvant therapy for immunotherapy rather than as a possible preventive strategy against the development of clinical symptoms in allergic patients [80].

5.2. Management of Peach Allergy

Once the diagnosis of peach allergy is made, peach should be eliminated from the patient's diet. Foods possibly cross-reacting with peach allergens should also be investigated by, firstly, assessing whether the patient is exposed to these foods without presenting symptoms and, if this is found not to be the case, by performing skin tests and/or specific IgE dosing. This strategy should mainly be considered for food cross-reacting via Pru p 3 (LTP) or Pru p 7 (peamaclein), given the higher risk of severe reaction associated with sensitization to these allergens.

Management strategies should include both the management of acute accidental reactions and long-term avoidance strategies.

A written emergency action plan for acute reactions should always be provided to all patients with peach allergy. In addition, two adrenaline auto-injectors (AAI) should be prescribed to patients with a history of anaphylaxis to peach.

In order to properly avoid peach, patients should also be educated on how to recognize the presence of peach in commercial products (such as fruit juices). Unfortunately, current labeling practices and legislation do not include the obligation to include the presence of this food, nor to highlight it on the label [81], which could result in the threat of accidental exposure. Other important aspects of educational programs for peach allergy include understanding and recognizing the early signs/symptoms of a possible allergic reaction, avoiding possible triggers or cofactors able to elicit the allergic reaction (e.g., asthma exacerbations, infections, exercise, alcohol, tiredness, use of NSAIDs, and menstruation), and knowing when and how to administer proper treatment, especially if an adrenaline auto-injector has been prescribed [34,40,78].

5.3. Allergen Immunotherapy

Immunotherapy is considered an attractive option to treat food allergies and aims at inducing immunological tolerance (the possibility of safe consumption, regardless of regular exposure) of foods.

In terms of oral immunotherapy (OIT), increasing amounts of food are administered to patients with a proven allergy in order to induce desensitization and, possibly, tolerance. In a paper by Patriarca et al. [82], one adult patient underwent OIT with peach and was successfully treated over a 3-month period. Nevertheless, the authors did not provide more specific details on this patient. A more recent study proposed a protocol using peach juice in 24 peach-allergic patients; the protocol followed a sublingual immunotherapy (SLIT) strategy [83]. At the end of the study, the authors were able to administer 200 mL of peach juice to 70.8% of their patients, without reporting severe adverse reactions during the challenge [83]. In any case, peach, as a wholly allergenic source, has not been an allergen on which researchers have focused their attention, as has been the case with OIT. Other SLIT protocols using specific peach proteins have been proposed, such as Pru p 3, on which several studies have been conducted. In the first published study on this topic, after 6 months of SLIT with peach extract quantified in mass units for Pru p 3, 33 patients showed an increase of 3-to-9 fold in their eliciting dose, with a significant difference when compared with the placebo group; moreover, no serious adverse events were reported, and the patients mainly experienced local reactions [84]. A more recent paper confirmed these results on 15 patients, even with an ultra-rush protocol [85]. Furthermore, Beitia et al. showed the effectiveness of Pru p 3 SLIT in a real-life study, including 29 patients, showing that, one year after starting SLIT, 73% had a negative challenge to peach, and, after 2 years, 95% of them did not react to the fruit [86]. In this study, the possibility of using Pru p 3 SLIT to treat patients suffering from LTP syndrome was confirmed, as also shown in other papers, with a positive impact on patients' quality of life as well [87,88]. Indeed, in the paper by González-Pérez et al., the authors showed that, in 18 adult patients treated for 3 consecutive years with Pru p 3 SLIT, the results on the Food Allergy Quality of Life Questionnaire-Adult Form (FAQLQ-AF) significantly decreased, showing a favorable impact on the patients' quality of life [87].

Finally, for patients suffering from PFAS, some authors focused on the possibility of treatment with subcutaneous immunotherapy (SCIT), using birch pollen extract. Nevertheless, researchers showed controversial results on this specific issue [89–91], and no study was specifically conducted on peach-allergic patients.

In general, even though peach OIT is possibly administered in research and specialized settings, there are currently insufficient data to be able to recommend this approach to treating patients in clinical practice [92].

6. Conclusions

Peach allergy may manifest with different clinical symptoms of ranging severity. Based on patients' sensitization profiles, clinicians may be able to highlight which patients are more at risk of developing a severe allergic reaction. Unfortunately, in clinical practice, clinicians are only able to dose specific serum IgE for the whole peach source, and for Pru p 1 (PR-10), Pru p 3 (LTP), Pru p 4 (profiline), and Pru p 7 (peamaclein). Patients monosensitized to Pru p 9 are known to be at risk of respiratory symptoms, while patients allergic to Pru p 3 and/or Pru p 7 are at risk of experiencing severe allergic reactions. To properly diagnose a peach allergy, therefore, besides presenting a compatible clinical history related to the fruit, patients need to have positive SPT and/or sIgE to available allergens. In cases in which diagnosis cannot be reached by combining these tests, an OFC may be performed, as this procedure is still considered the gold diagnostic standard. Once the diagnosis is made, patients and caregivers should receive proper education on peach avoidance strategies, an emergency action plan for accidental acute reactions and AAIs in case of history of anaphylaxis to peach. OIT is a promising treatment for patients with food allergies who are at high risk of a life-threatening reaction or severe impairment of quality of life. However, currently, peach immunotherapy is not advised in clinical practice.

Author Contributions: Conceptualization, S.B., D.C., F.C., P.C., C.M., U.P., F.P., G.L.M. and S.A.; data curation, S.B., D.C., F.C., P.C., C.M. and S.A.; writing—original draft preparation, S.B., D.C., F.C., P.C., C.M. and S.A.; writing—review and editing, S.B., D.C., F.C., P.C., C.M., U.P., F.P., G.L.M. and S.A.; supervision, S.B., D.C., F.C., P.C., C.M., U.P., F.P., G.L.M. and S.A. All authors have read and agreed to the published version of the manuscript.

Funding: The publication fee was financed by the Italian Society of Pediatric Allergy and Immunology (Societa' italiana di allergologia e Immunologia pediatrica, SIAIP, Via Libero Temolo 4, 20126 Milano). However, no significant funding source could have influenced the outcomes of this work.

Institutional Review Board Statement: Not applicable.

Informed Consent Statement: Not applicable.

Data Availability Statement: Not applicable.

Acknowledgments: The authors want to thank the Italian Society of Pediatric Allergy and Immunology (Societa' italiana di allergologia e Immunologia pediatrica, SIAIP, Via Libero Temolo 4, 20126 Milano) for its support in relation to this work.

Conflicts of Interest: The authors declare no conflict of interest.

References

1. Haleema, R.; Shenoy, A.; Shabraya, A.R. A Review on Pharmacological Activities of Prunus persica. *Int. J. Pharm. Sci. Rev. Res.* **2020**, *60*, 38–40.
2. Christenhusz, M.J.M.; Byng, J.W. The number of known plants species in the world and its annual increase. *Phytotaxa* **2016**, *261*, 201–217. [CrossRef]
3. The Plant List: Rosaceae. Available online: http://www.theplantlist.org/1.1/browse/A/Rosaceae (accessed on 3 January 2022).
4. The Families of Flowering Plants. Available online: http://www1.biologie.uni-hamburg.de/b-online/delta/angio/index.htm (accessed on 3 January 2022).
5. Kant, R.; Shukla, K.R.; Shukla, A. A Review on Peach (Prunus persica): An Asset of Medicinal Phytochemicals. *Int. J. Res. Appl. Sci. Eng. Technol.* **2018**, *6*, 2186–2200. [CrossRef]
6. Cuesta-Herranz, J.; Lázaro, M.; de las Heras, M.; Lluch, M.; Figueredo, E.; Umpierrez, A.; Hernandez, J.; Cuesta, C. Peach allergy pattern: Experience in 70 patients. *Allergy* **1998**, *53*, 78–82. [CrossRef] [PubMed]
7. Zuidmeer, L.; Goldhahn, K.; Rona, R.J.; Gislason, D.; Madsen, C.; Summers, C.; Sodergren, E.; Dahlstrom, J.; Lindner, T.; Sigurdardottir, S.T.; et al. The prevalence of plant food allergies: A systematic review. *J. Allergy Clin. Immunol.* **2008**, *121*, 1210–1218. [CrossRef] [PubMed]
8. Burney, P.; Summers, C.; Chinn, S.; Hooper, R.; van Ree, R.; Lidholm, J. Prevalence and distribution of sensitization to foods in the European Community Respiratory Health Survey: A EuroPrevall analysis. *Allergy* **2010**, *65*, 1182–1188. [CrossRef] [PubMed]
9. Burney, P.G.; Potts, J.; Kummeling, I.; Mills, E.N.C.; Clausen, M.; Dubakiene, R.; Barreales, L.; Fernandez-Perez, C.; Fernandez-Rivas, M.; Le, T.M.; et al. The prevalence and distribution of food sensitization in European adults. *Allergy* **2014**, *69*, 365–371. [CrossRef] [PubMed]

10. Boyano-Martinez, T.; Pedrosa, M.; Belver, T.; Quirce, S.; Garcia-Ara, C. Peach allergy in Spanish children: Tolerance to the pulp and molecular sensitization profile. *Pediatr. Allergy Immunol.* **2013**, *24*, 168–172. [CrossRef] [PubMed]

11. Pastorello, E.A.; Farioli, L.; Stafylaraki, C.; Mascheri, A.; Scibilla, J.; Pravettoni, V.; Primavesi, L.; Piantanida, M.; Nichelatti, M.; Asero, R. Anti-rPru p 3 IgE levels are inversely related to the age at onset of peach-induced severe symptoms reported by peach allergic adults. *Int. Arch. Allergy Immunol.* **2013**, *162*, 45–49. [CrossRef] [PubMed]

12. Le, T.M.; Lindner, T.M.; Pasmans, S.G.; Guikers, C.L.H.; van Hoffen, E.; Bruijnzeel-Koomen, C.A.F.M.; Knulst, A.C. Reported food allergy to peanut, tree nuts and fruit: Comparison of clinical manifestations, prescription of medication and impact on daily life. *Allergy* **2008**, *63*, 910–916. [CrossRef]

13. Allergen Nomenclature. WHO/IUIS Allergne Nomenclature Sub-Committee. Available online: http://www.allergen.org/search.php?allergensource=Prunus+persica (accessed on 3 January 2022).

14. Gaier, S.; Marsh, J.; Oberhuber, C.; Rigby, N.M.; Lovegrove, A.; Alessandri, S.; Briza, P.; Radauer, C.; Zuidmeer, L.; van Ree, R.; et al. Purification and structural stability of the peach allergens Pru p 1 and Pru p 3. *Mol. Nutr. Food Res.* **2008**, *52* (Suppl. S2), S220–S229. [CrossRef]

15. Gaier, S.; Oberhuber, C.; Hemmer, W.; Radauer, C.; Rigby, N.M.; Marsh, J.T.; Mills, C.E.; Shewry, P.R.; Hoffmann-Sommergruber, K. Pru p 3 as a marker for symptom severity for patients with peach allergy in a birch pollen environment. *J. Allergy Clin. Immunol.* **2009**, *124*, 166–167. [CrossRef] [PubMed]

16. Sastre, J. Molecular diagnosis in allergy. *Clin. Exp. Allergy* **2010**, *40*, 1442–1460. [CrossRef]

17. Carnes, J.; Fernandez-Caldas, E.; Gallego, M.T.; Ferrer, A.; Cuesta-Herranz, J. Pru p 3 (LTP) content in peach extracts. *Allergy* **2002**, *57*, 1071–1075. [CrossRef] [PubMed]

18. Peterson, A.; Kleine-Tebbe, J.; Scheurer, S. Stable Plant Food Allergens I: Lipid Transfer Proteins. In *Molecular Allergy Diagnostics*, 1st ed.; Kleine-Tebbe, J., Jacob, T., Eds.; Springer: Cham, Switzerland, 2017; pp. 57–71. [CrossRef]

19. Egger, M.; Hauser, M.; Mari, A.; Ferreira, F.; Gadermaier, G. The role of lipid transfer proteins in allergic diseases. *Curr. Allergy Asthma Rep.* **2010**, *10*, 326–335. [CrossRef]

20. Pascal, M.; Munoz-Cano, R.; Reina, Z.; Palacin, A.; Vilella, R.; Picado, C.; Juan, M.; Sánchez-López, J.; Rueda, M.; Salcedo, G.; et al. Lipid transfer protein syndrome: Clinical pattern, cofactor effect and profile of molecular sensitization to plant-foods and pollens. *Clin. Exp. Allergy* **2012**, *42*, 1529–1539. [CrossRef] [PubMed]

21. Zamieskova, L.; Žiarovská, J.; Bilčíková, J.; Fialková, V. Natural variability of restriction profiles in non-coding part of *Prunus persica* (L.) Batsch. Pru p 3 gene. *Acta Fytotech. Zootech.* **2020**, *23*, 1–6. [CrossRef]

22. Salcedo, G.; Sanchez-Monge, R.; Barber, D.; Diaz-Perales, A. Plant non-specific lipid transfer proteins: An interface between plant defence and human allergy. *Biochim. Biophys. Acta* **2007**, *1771*, 781–791. [CrossRef] [PubMed]

23. Cuesta-Herranz, J.; Lazaro, M.; Martinez, A.; Figueredo, E.; Palacios, R.; de-Las-Heras, M.; Martínez, J. Pollen allergy in peach-allergic patients: Sensitization and cross-reactivity to taxonomically unrelated pollens. *J. Allergy Clin. Immunol.* **1999**, *104 Pt 1*, 688–694. [CrossRef]

24. Ando, Y.; Miyamoto, M.; Kato, M.; Nakayama, M.; Fukuda, H.; Yoshihara, S. Pru p 7 Predicts Severe Reactions after Ingestion of Peach in Japanese Children and Adolescents. *Int. Arch. Allergy Immunol.* **2020**, *181*, 183–190. [CrossRef] [PubMed]

25. Yang, Z.; Ma, Y.; Chen, L.; Xie, R.; Zhang, X.; Zhang, B.; Lu, M.; Wu, S.; Gilissen, L.J.W.J.; van Ree, R.; et al. Differential transcript abundance and genotypic variation of four putative allergen-encoding gene families in melting peach. *Tree Genet. Genom.* **2011**, *7*, 903–916. [CrossRef]

26. Pastorello, E.A.; Farioli, L.; Pravettoni, V.; Scibilia, J.; Mascheri, A.; Borgonovo, L.; Piantanida, M.; Primavesi, L.; Stafylaraki, C.; Pasqualetti, S.; et al. Pru p 3-sensitised Italian peach-allergic patients are less likely to develop severe symptoms when also presenting IgE antibodies to Pru p 1 and Pru p 4. *Int. Arch. Allergy Immunol.* **2011**, *156*, 362–372. [CrossRef] [PubMed]

27. Tuppo, L.; Alessandri, C.; Pomponi, D.; Picone, D.; Tamburini, M.; Ferrara, R.; Petriccione, M.; Mangone, I.; Palazzo, P.; Liso, M.; et al. Peamaclein—A new peach allergenic protein: Similarities, differences and misleading features compared to Pru p 3. *Clin. Exp. Allergy* **2013**, *43*, 128–140. [CrossRef] [PubMed]

28. Nahirnak, V.; Rivarola, M.; Almasia, N.I.; Barrios Baron, M.P.; Hopp, H.E.; Vile, D.; Vile, D.; Paniego, N.; Vazquez Rovere, C. *Snakin-1* affects reactive oxygen species and ascorbic acid levels and hormone balance in potato. *PLoS ONE* **2019**, *14*, e0214165. [CrossRef]

29. Inomata, N.; Miyakawa, M.; Aihara, M. Gibberellin-regulated protein in Japanese apricot is an allergen cross-reactive to Pru p 7. *Immun. Inflamm. Dis.* **2017**, *5*, 469–479. [CrossRef] [PubMed]

30. Tuppo, L.; Alessandri, C.; Pasquariello, M.S.; Petriccione, M.; Giangrieco, I.; Tamburrini, M.; Mari, A.; Ciardiello, M.A. Pomegranate Cultivars: Identification of the New IgE-Binding Protein Pommaclein and Analysis of Antioxidant Variability. *J. Agric. Food Chem.* **2017**, *65*, 2702–2710. [CrossRef]

31. Inomata, N.; Miyakawa, M.; Ikeda, N.; Oda, K.; Aihara, M. Identification of gibberellin-regulated protein as a new allergen in orange allergy. *Clin. Exp. Allergy* **2018**, *48*, 1509–1520. [CrossRef]

32. Charpin, D.; Pichot, C.; Belmonte, J.; Sutra, J.; Zidkova, J.; Chanez, P.; Shahali, Y.; Sénéchal, H.; Poncet, P. Cypress Pollinosis: From Tree to Clinic. *Clin. Rev. Allergy Immunol.* **2019**, *56*, 174–195. [CrossRef] [PubMed]

33. Klingebiel, C.; Chantran, Y.; Arif-Lusson, R.; Ehrenberg, A.E.; Östling, J.; Poisson, A.; Liabeuf, V.; Agabriel, C.; Birnbaum, J.; Porri, F.; et al. Pru p 7 sensitization is a predominant cause of severe, cypress pollen-associated peach allergy. *Clin. Exp. Allergy* **2019**, *49*, 526–536. [CrossRef]

34. Biagioni, B.; Tomei, L.; Valleriani, C.; Liccioli, G.; Barni, S.; Sarti, L.; Citera, F.; Giovannini, M.; Mori, F. Allergy to Gibberellin-Regulated Proteins (Peamaclein) in Children. *Int. Arch. Allergy Immunol.* **2021**, *182*, 1194–1199. [CrossRef]

35. Allergy & Autoimmune Disease. f95 Peach. Available online: https://www.thermofisher.com/diagnostic-education/hcp/it/it/resource-center/allergen-encyclopedia/whole-allergens.html?key=f95 (accessed on 3 January 2022).

36. Palacin, A.; Rivas, L.A.; Gomez-Casado, C.; Aguirre, J.; Tordesillas, L.; Bartra, J.; Blanco, C.; Carrillo, T.; Cuesta-Herranz, J.; Bonny, J.A.; et al. The involvement of thaumatin-like proteins in plant food cross-reactivity: A multicenter study using a specific protein microarray. *PLoS ONE* **2012**, *7*, e44088. [CrossRef] [PubMed]

37. González-Mancebo, E.; Gonzalez-de-Olano, D.; Trujillo, M.; Santos, S.; Gandolfo-Cano, M.; Melendez, A.; Juárez, R.; Morales, P.; Calso, A.; Mazuela, O.; et al. Prevalence of Sensitization to Lipid Transfer Proteins and Profilins in a Population of 430 Patients in the South of Madrid. *J. Investig. Allergol. Clin. Immunol.* **2011**, *21*, 278–282. [PubMed]

38. Victorio Puche, L.; Somoza, M.L.; Lopez-Sanchez, J.D.; Garrido-Arandia, M.; Diaz-Peralesm, A.; Blanca, M. Peach Tree Pollen and Prunus persica 9 Sensitisation and Allergy in Children and Adolescents. *Int. Arch. Allergy Immunol.* **2019**, *180*, 212–220. [CrossRef] [PubMed]

39. Somoza, M.L.; Pérez-Sánchez, N.; Victorio-Puche, L.; Martín-Pedraza, L.; Esteban Rodríguez, A.; Blanca-López, N.; Abel Fernández González, E.; Ruano-Zaragoza, M.; Prieto-Moreno Pfeifer, A.; Fernández Caldas, E.; et al. Subjects develop tolerance to Pru p 3 but respiratory allergy to Pru p 9: A large study group from a peach exposed population. *PLoS ONE* **2021**, *16*, e0255305. [CrossRef] [PubMed]

40. Muraro, A.; Werfel, T.; Hoffmann-Sommergruber, K.; Roberts, G.; Beyer, K.; Bindslev-Jensen, C.; Cardona, V.; Dubois, A.; duToit, G.; Eigenmann, P.; et al. EAACI food allergy and anaphylaxis guidelines: Diagnosis and management of food allergy. *Allergy* **2014**, *69*, 1008–1025. [CrossRef] [PubMed]

41. Asero, R. Peach-induced contact urticaria is associated with lipid transfer protein sensitization. *Int. Arch. Allergy Immunol.* **2011**, *154*, 345–348. [CrossRef] [PubMed]

42. Pérez-Calderón, R.; Gonzalo-Garijo, M.Á.; Rodríguez-Velasco, F.J.; Sánchez-Vega, S.; Bartolomé-Zavala, B. Occupational respiratory allergy in peach crop workers. *Allergy* **2017**, *72*, 1556–1564. [CrossRef] [PubMed]

43. Fernandez-Rivas, M.; Bolhaar, S.; González-Mancebo, E.; Asero, R.; van Leeuwen, A.; Bohle, B.; Ma, Y.; Ebner, C.; Rigby, N.; Sancho, A.I.; et al. Apple allergy across Europe: How allergen sensitization profiles determine the clinical expression of allergies to plant foods. *J. Allergy Clin. Immunol.* **2006**, *118*, 481–488. [CrossRef] [PubMed]

44. Fernandez-Rivas, M.; González-Mancebo, E.; Rodríguez-Pérez, R.; Benito, C.; Sánchez-Monge, R.; Salcedo, G.; Alonso, M.D.; Rosado, A.; Tejedor, M.A.; Vila, C.; et al. Clinically relevant peach allergy is related to peach lipid transfer protein, Pru p 3, in the Spanish population. *J. Allergy Clin. Immunol.* **2003**, *112*, 789–795. [CrossRef]

45. Lee, W.J.; Kim, D.H.; Woo, S.H.; Seol, S.H.; Choi, S.P. Targeted temperature management after cardiac arrest with anaphylaxis. *Am. J. Emerg. Med.* **2017**, *35*, 807. [CrossRef] [PubMed]

46. Barradas Lopes, J.; Santa, C.; Valente, C.; Presa, A.R.; Sousa, M.J.; Reis Ferreira, A. Allergy to lipid transfer proteins (LTP) in a pediatric population. *Eur. Ann. Allergy Clin. Immunol.* **2021**. [CrossRef] [PubMed]

47. Poncet, P.; Sénéchal, H.; Charpin, D. Update on pollen-food allergy syndrome. *Expert. Rev. Clin. Immunol.* **2020**, *16*, 561–578. [CrossRef] [PubMed]

48. Werfel, T.; Asero, R.; Ballmer-Weber, B.K.; Beyer, K.; Enrique, E.; Knulst, A.C.; Mari, A.; Muraro, A.; Ollert, M.; Poulsen, L.K.; et al. Position paper of the EAACI: Food allergy due to immunological cross-reactions with common inhalant allergens. *Allergy* **2015**, *70*, 1079–1090. [CrossRef] [PubMed]

49. Ma, S.; Sicherer, S.H.; Nowak-Wegrzyn, A. A survey on the management of pollen-food allergy syndrome in allergy practices. *J. Allergy Clin. Immunol.* **2003**, *112*, 784–788. [CrossRef]

50. Kim, M.; Ahn, Y.; Yoo, Y.; Kim, D.K.; Yang, H.J.; Park, H.S.; Lee, H.J.; Kim, M.A.; Jeong, Y.Y.; Kim, B.S.; et al. Clinical Manifestations and Risk Factors of Anaphylaxis in Pollen-Food Allergy Syndrome. *Yonsei Med. J.* **2019**, *60*, 960–968. [CrossRef] [PubMed]

51. Sicherer, S.H.; Sampson, H.A. Food allergy. *J. Allergy Clin. Immunol.* **2010**, *125* (Suppl. S2), S116–S125. [CrossRef]

52. Sanchez-López, J.; Tordesillas, L.; Pascal, M.; Muñoz-Cano, R.; Garrido, M.; Rueda, M.; Vilella, R.; Valero, A.; Díaz-Perales, A.; Picado, C.; et al. Role of Art v 3 in pollinosis of patients allergic to Pru p 3. *J. Allergy Clin. Immunol* **2014**, *133*, 1018–1025. [CrossRef] [PubMed]

53. Borghesan, F.; Mistrello, G.; Roncarolo, D.; Amato, S.; Plebani, M.; Asero, R. Respiratory allergy to lipid transfer protein. *Int. Arch. Allergy Immunol.* **2008**, *147*, 161–165. [CrossRef]

54. Lyons, S.A.; Burney, P.G.J.; Ballmer-Weber, B.K.; Fernandez-Rivas, M.; Barreales, L.; Clausen, M.; Dubakiene, R.; Fernandez-Perez, C.; Fritsche, P.; Jedrzejczak-Czechowicz, M.; et al. Food Allergy in Adults: Substantial Variation in Prevalence and Causative Foods Across Europe. *J. Allergy Clin. Immunol. Pract.* **2019**, *7*, 1920–1928.e11. [CrossRef] [PubMed]

55. Asero, R.; Antonicelli, L.; Arena, A.; Bommarito, L.; Caruso, B.; Colombo, G.; Crivellaro, M.; De Carli, M.; Della Torre, E.; Della Torre, F.; et al. Causes of food-induced anaphylaxis in Italian adults: A multi-centre study. *Int. Arch. Allergy Immunol.* **2009**, *150*, 271–277. [CrossRef] [PubMed]

56. Skypala, I.J.; Bartra, J.; Ebo, D.G.; Antje Faber, M.; Fernández-Rivas, M.; Gomez, F.; Luengo, O.; Till, S.J.; Asero, R.; Barber, D.; et al. The diagnosis and management of allergic reactions in patients sensitized to non-specific lipid transfer proteins. *Allergy* **2021**, *76*, 2433–2446. [CrossRef] [PubMed]

57. Asero, R.; Piantanida, M.; Pinter, E.; Pravettoni, V. The clinical relevance of lipid transfer protein. *Clin. Exp. Allergy* **2018**, *48*, 6–12. [CrossRef] [PubMed]

58. Scala, E.; Till, S.J.; Asero, R.; Abeni, D.; Guerra, E.C.; Pirrotta, L.; Paganelli, R.; Pomponi, D.; Giani, M.; De Pità, O.; et al. Lipid transfer protein sensitization: Reactivity profiles and clinical risk assessment in an Italian cohort. *Allergy* **2015**, *70*, 933–943. [CrossRef] [PubMed]

59. Basagaña, M.; Elduque, C.; Teniente-Serra, A.; Casas, I.; Roger, A. Clinical Profile of Lipid Transfer Protein Syndrome in a Mediterranean Area. *J. Investig. Allergol. Clin. Immunol.* **2018**, *28*, 58–60. [CrossRef] [PubMed]

60. Bogas, G.; Muñoz-Cano, R.; Mayorga, C.; Casas, R.; Bartra, J.; Pérez, N.; Pascal, M.; Palomares, F.; Torres, M.J.; Gómez, F. Phenotyping peach-allergic patients sensitized to lipid transfer protein and analysing severity biomarkers. *Allergy* **2020**, *75*, 3228–3236. [CrossRef]

61. NCBI, National Center for Biotechnology Information, Bethesda, USA. Basic Local Alignment Search Tool. Available online: http://blast.ncbi.nlm.nih.gov (accessed on 10 February 2022).

62. Matricardi, P.M.; Salcedo, G.; Sanchez-Monge, R.; Diaz-Perales, A.; Garcia-Casado, G.; Barber, D. Plant non-specific lipid transfer proteins as food and pollen allergens. *Clin. Exp. Allergy* **2004**, *34*, 1336–1341.

63. Ukleja-Sokołowska, N.; Zacniewski, R.; Gawrońska-Ukleja, E.; Żbikowska-Gotz, M.; Lis, K.; Sokołowski, Ł.; Adamczak, R.; Bartuzi, Z. Food-dependent, exercise-induced anaphylaxis in a patient allergic to peach. *Int. J. Immunopathol. Pharmacol.* **2018**, *32*, 2058738418803154. [CrossRef]

64. Romano, A.; Scala, E.; Rumi, G.; Gaeta, F.; Caruso, C.; Alonzi, C.; Maggioletti, M.; Ferrara, R.; Palazzo, P.; Palmieri, V.; et al. Lipid transfer proteins: The most frequent sensitizer in Italian subjects with food-dependent exercise-induced anaphylaxis. *Clin. Exp. Allergy* **2012**, *42*, 1643–1653. [CrossRef] [PubMed]

65. Sénéchal, H.; Santrucek, J.; Melcova, M.; Svoboda, P.; Zídková, J.; Charpin, D.; Guilloux, L.; Shahali, Y.; Selva, M.A.; Couderc, R.; et al. A new allergen family involved in pollen food associated syndrome: Snakin/gibberellin regulated proteins. *J. Allergy Clin. Immunol.* **2018**, *141*, 411–414. [CrossRef] [PubMed]

66. Inomata, N. Gibberellin-regulated protein allergy: Clinical features and cross reactivity. *Allergol. Int.* **2020**, *69*, 11–18. [CrossRef]

67. Tuppo, L.; Alessandri, C.; Giangrieco, I.; Tamburrini, M.; Arriaza, R.H.; Chruszcz, M.; Mari, A.; Ciardiello, M.A. When the Frequencies of Sensitization and Elicitation of Allergic Reaction Do Not Correlate—The Case of Ap-ple Gibberellin-Regulated Protein Tested in an Italian Population. *Front. Allergy* **2021**, *2*, 745825. [CrossRef]

68. Muraro, A.; Fernandez-Rivas, M.; Beyer, K.; Cardona, V.; Clark, A.; Eller, E.; Hourihane, J.O.B.; Jutel, M.; Sheikh, A.; Agache, I.; et al. The urgent need for a harmonized severity scoring system for acute allergic reactions. *Allergy* **2018**, *73*, 1792–1800. [CrossRef] [PubMed]

69. Inomata, N.; Miyakawa, M.; Aihara, M. Eyelid edema as a predictive factor for sensitization to Pru p 7 in peach allergy. *J. Dermatol.* **2016**, *43*, 900–905. [CrossRef]

70. Burks, A.W.; Tang, M.; Sicherer, S.; Muraro, A.; Eigenmann, P.A.; Ebisawa, M.; Fiocchi, A.; Chiang, W.; Beyer, K.; Wood, R.; et al. ICON: Food allergy. *J. Allergy Clin. Immunol.* **2012**, *129*, 906–920. [CrossRef]

71. Inomata, N.; Okazaki, F.; Moriyama, T.; Nomura, Y.; Yamaguchi, Y.; Honjoh, T.; Kawamura, Y.; Narita, H.; Aihara, M. Identification of peamaclein as a marker allergen related to systemic reactions in peach allergy. *Ann. Allergy Asthma Immunol.* **2014**, *112*, 175–177. [CrossRef] [PubMed]

72. Matricardi, P.M.; Kleine-Tebbe, J.; Hoffmann, H.J.; Valenta, R.; Hilger, C.; Hofmaier, S.; Aalberse, R.C.; Agache, I.; Asero, R.; Ballmer-Weber, B.; et al. EAACI Molecular Allergology User's Guide. *Pediatr. Allergy Immunol.* **2016**, *27* (Suppl. S23), 1–250. [CrossRef] [PubMed]

73. Muraro, A.; Lemanske, R.F.; Castells, M.; Torres, M.J.; Khan, D.; Simon, H.-U.; Bindslev-Jensen, C.; Burks, W.; Poulsen, L.K.; Sampson, H.A.; et al. Precision medicine in allergic disease-food allergy, drug allergy, and anaphylaxis-PRACTALL document of the European Academy of Allergy and Clinical Immunology and the American Academy of Allergy, Asthma and Immunology. *Allergy* **2017**, *72*, 1006–1021. [CrossRef]

74. Asero, R.; Aruanno, A.; Bresciani, M.; Brusca, I.; Carollo, M.; Cecchi, L.; Cortellini, G.; Deleonardi, G.; Farsi, A.; Ferrarini, E.; et al. Evaluation of two commercial peach extracts for skin prick testing in the diagnosis of hypersensitivity to lipid transfer protein. A multicenter study. *Eur. Ann. Allergy Clin. Immunol.* **2021**, *53*, 168–170. [CrossRef]

75. Somoza, M.L.; Prieto-Moreno Pfeifer, A.; Martín-Pedraza, L.; Victorio Puche, L.; Esteban Rodríguez, A.; Blanca-López, N.; Eva Fernández González, A.; Fernández-Caldas, E.; Morán Morales, M.; Fernández-Sánchez-, F.J.; et al. Skin Testing With Peach Peel Extract *Versus Serum* IgE to Pru p 3 as a Stronger Predictor of Peach-Induced Anaphylaxis. *Allergy Asthma Immunol. Res.* **2021**, *13*, 922–932. [CrossRef] [PubMed]

76. Kennard, L.; Thomas, I.; Rutkowski, K.; Azzu, V.; Yong, P.F.K.; Kasternow, B.; Hunter, H.; Cabdi, N.M.O.; Nakonechna, A.; Wagner, A. A Multicenter Evaluation of Diagnosis and Management of Omega-5 Gliadin Allergy (Also Known as Wheat-Dependent Exercise-Induced Anaphylaxis) in 132 Adults. *J. Allergy Clin. Immunol. Pract.* **2018**, *6*, 1892–1897. [CrossRef]

77. O'Keefe, A.W.; De Schryver, S.; Mill, J.; Mill, C.; Dery, A.; Ben-Shosha, M. Diagnosis and management of food allergies: New and emerging options: A systematic review. *J. Asthma Allergy* **2014**, *7*, 141–164. [CrossRef] [PubMed]

78. Caimmi, D.; Caffarelli, C.; Licari, A.; Miraglia del Giudice, M.; Calvani, M.; Marseglia, G.L.; Marseglia, A.; Ricci, G.; Martelli, A.; Cravidi, C.; et al. Food allergy in primary care. *Acta Biomed.* **2021**, *92* (Suppl. S7), e2021521. [CrossRef]

79. Fleischer, D.M.; Chan, E.S.; Venter, C.; Spergel, J.M.; Abrams, E.M.; Stukus, D.; Groetch, M.; Shaker, M.; Greenhawt, M. A consensus approach to the primary prevention of food allergy through nutrition: Guidance from the American Academy of Allergy, Asthma, and Immunology; American College of Allergy, Asthma, and Immunology; and the Canadian Society for Allergy and Clinical Immunology. *J. Allergy Clin. Immunol. Pract.* **2021**, *9*, 22–43. [CrossRef]

80. Fiocchi, A.; Vickery, B.P.; Wood, R.A. The use of biologics in food allergy. *Clin. Exp. Allergy* **2021**, *51*, 1006–1018. [CrossRef]

81. Food Safety. EU Law on Food Information to Consumers. Available online: https://ec.europa.eu/food/safety/labelling-and-nutrition/food-information-consumers-legislation_it (accessed on 5 January 2022).

82. Patriarca, G.; Nucera, E.; Roncallo, C.; Pollastrini, E.; Bartolozzi, F.; De Pasquale, T.; Buonomo, A.; Gasbarrini, G.; Di Campli, C.; Schiavino, D. Oral desensitizing treatment in food allergy: Clinical and immunological results. *Aliment. Pharmacol. Ther.* **2003**, *17*, 459–465. [CrossRef] [PubMed]

83. Navarro, B.; Alarcon, E.; Claver, A.; Pascal, M.; Diaz-Perales, A.; Cistero-Bahima, A. Oral immunotherapy with peach juice in patients allergic to LTPs. *Allergy Asthma Clin. Immunol.* **2019**, *15*, 60. [CrossRef]

84. Fernández-Rivas, M.; Garrido Fernández, S.; Nadal, J.A.; Díaz de Durana, M.D.; García, B.E.; González-Mancebo, E.; Martín, S.; Barber, D.; Rico, P.; Tabar, A.I. Randomized double-blind, placebo-controlled trial of sublingual immunotherapy with a Pru p 3 quantified peach extract. *Allergy* **2009**, *64*, 876–883. [CrossRef]

85. Moura, A.L.; Pereira, C.; Regateiro, F.S.; Azevedo, J.; Todo Bom, A.; Carrapatoso, I. Pru p 3 sublingual immunotherapy ultra-rush protocol is safe and clinically effective. *Eur. Ann. Allergy Clin. Immunol.* **2019**, *51*, 206–212. [CrossRef]

86. Beitia, J.M.; Castro, A.V.; Cardenas, R.; Pena-Arellano, M.I. Pru p 3 sublingual immunotherapy in patients with Lipid Tranfer Protein Syndrome: Is it worth? *Int. Arch. Allergy Immunol.* **2021**, *182*, 447–454. [CrossRef] [PubMed]

87. González Pérez, A.; Carbonell Martínez, A.; Escudero Pastor, A.I.; Navarro Garrido, C.; Miralles López, J.C. Pru p 3 oral immunotherapy efficacy, induced immunological changes and quality of life improvement in patients with LTP syndrome. *Clin. Transl. Allergy* **2020**, *10*, 20. [CrossRef] [PubMed]

88. García-Gutiérrez, I.; Medellín, D.R.; Noguerado-Mellado, B.; Ordoñez, C.L.; Abreu, M.G.; Jimeno Nogales, L.; Rojas-Pérez-Ezquerra, P. Treatment with lipid transfer protein sublingual immunotherapy: Slowing down new sensitizations. *Asia Pac. Allergy* **2021**, *11*, e6. [CrossRef] [PubMed]

89. Caimmi, D.; Demoly, P. Guidelines for the prescription of allergen immunotherapy and patient's follow-up—Clinical questions and revision of the literature. *Rev. Fr. Allergol.* **2021**, *61*, 35–56. [CrossRef]

90. Hamada, M.; Kagawa, M.; Tanaka, I. Evaluation of subcutaneous immunotherapy with birch pollen extract for pollen-food allergy syndrome. *Asia Pac. Allergy* **2021**, *11*, e39. [CrossRef]

91. Van der Valk, J.P.M.; Nagl, B.; van Wljk, R.G.; Bohle, B.; de Jong, N.W. The Effect of Birch Pollen Immunotherapy on Apple and rMal d 1 Challenges in Adults with Apple Allergy. *Nutrients* **2020**, *12*, 519. [CrossRef] [PubMed]

92. Pajno, G.B.; Fernandez-Rivas, M.; Arasi, S.; Roberts, G.; Akdis, C.A.; Alvaro-Lozano, M.; Beyer, K.; Bindslev-Jensen, C.; Burks, W.; Ebisawa, M.; et al. EAACI Guidelines on allergen immunotherapy: IgE-mediated food allergy. *Allergy* **2018**, *73*, 799–815. [CrossRef]

Article

Threshold of Reactivity and Tolerance to Precautionary Allergen-Labelled Biscuits of Baked Milk- and Egg-Allergic Children

Vincenzo Fierro [1,†], Valeria Marzano [2,†], Linda Monaci [3], Pamela Vernocchi [2], Maurizio Mennini [1], Rocco Valluzzi [1], Stefano Levi Mortera [2], Rosa Pilolli [3], Lamia Dahdah [1], Veronica Calandrelli [1], Giorgia Bracaglia [4], Stefania Arasi [1], Carla Riccardi [1], Alessandro Fiocchi [1,*] and Lorenza Putignani [5]

[1] Translational Research in Pediatric Specialities Area, Allergy Unit, Bambino Gesù Children's Hospital, IRCCS, 0165 Rome, Italy; vincenzo.fierro@opbg.net (V.F.); maurizio.mennini@opbg.net (M.M.); roccoluigi.valluzzi@opbg.net (R.V.); lamiaantanios.dahdah@opbg.net (L.D.); veronica.calandrelli@opbg.net (V.C.); stefania.arasi@opbg.net (S.A.); carla.riccardi@opbg.net (C.R.)

[2] Multimodal Laboratory Medicine Research Area, Unit of Human Microbiome, Bambino Gesù Children's Hospital, IRCCS, 00165 Rome, Italy; valeria.marzano@opbg.net (V.M.); pamela.vernocchi@opbg.net (P.V.); stefano.levimortera@opbg.net (S.L.M.)

[3] Institute of Sciences of Food Production, CNR-ISPA, 70126 Bari, Italy; linda.monaci@ispa.cnr.it (L.M.); rosa.pilolli@ispa.cnr.it (R.P.)

[4] Department of Diagnostics and Laboratory Medicine, Unit of Allergy and Autoimmunity, Bambino Gesù Children's Hospital, IRCCS, 00165 Rome, Italy; giorgia.bracaglia@opbg.net

[5] Department of Diagnostics and Laboratory Medicine, Unit of Microbiology and Diagnostic Immunology, Unit of Microbiomics, and Multimodal Laboratory Medicine Research Area, Unit of Human Microbiome, Bambino Gesù Children's Hospital, IRCCS, 00165 Rome, Italy; lorenza.putignani@opbg.net

* Correspondence: agiovanni.fiocchi@opbg.net; Tel.: +39-0668-594777

† These authors contributed equally to this work.

Citation: Fierro, V.; Marzano, V.; Monaci, L.; Vernocchi, P.; Mennini, M.; Valluzzi, R.; Levi Mortera, S.; Pilolli, R.; Dahdah, L.; Calandrelli, V.; et al. Threshold of Reactivity and Tolerance to Precautionary Allergen-Labelled Biscuits of Baked Milk- and Egg-Allergic Children. *Nutrients* **2021**, *13*, 4540. https://doi.org/10.3390/nu13124540

Academic Editor: Eva Untersmayr

Received: 18 November 2021
Accepted: 16 December 2021
Published: 18 December 2021

Publisher's Note: MDPI stays neutral with regard to jurisdictional claims in published maps and institutional affiliations.

Abstract: Extremely sensitive food-allergic patients may react to very small amounts of allergenic foods. Precautionary allergen labelling (PAL) warns from possible allergenic contaminations. We evaluated by oral food challenge the reactivity to a brand of PAL-labelled milk- and egg-free biscuits of children with severe milk and egg allergy. We explored the ability of proteomic methods to identify minute amounts of milk/egg allergens in such biscuits. Traces of milk and/or egg allergens in biscuits were measured by two different liquid-chromatography-mass spectrometry methods. The binding of patient's serum with egg/milk proteins was assessed using immunoblotting. None of the patients reacted to biscuits. Egg and milk proteins were undetectable with a limit of detection of 0.6 µg/g for milk and egg (method A), and of 0.1 and 0.3 µg /g for milk and egg, respectively (method B). The immunoblots did not show milk/egg proteins in the studied biscuits. Milk/egg content of the biscuits is far lower than 4 µg of milk or egg protein per gram of product, the minimal doses considered theoretically capable of causing reactions. With high sensitivity, proteomic assessments predict the harmlessness of very small amount of allergens in foods, and can be used to help avoiding unnecessary PAL.

Keywords: labelling; food allergy; prevention; proteomics; mass spectrometry; cow's milk allergy

1. Introduction

The subset of food-allergic patients sensitive to minute amounts of foods is facing problems of food safety every day [1]. To protect them from accidental ingestions, regulatory authorities have put in place legislative measures prescribing the declaration of food allergen ingredients in the respective food labels [2]. Beyond food allergen ingredients, precautionary labelling of allergens (PAL) has been adopted by food producers as additional level of protection when food allergens may contaminate foods. PAL conveys an "information on the possible and unintentional presence in food of substances or products causing allergies or intolerances, provided voluntarily by the food business operator" [3].

PAL may further reduce the possible food choices of consumers who are already forced to reduce their options [4]. Conversely, PAL-free foods may contain significant amounts of food allergens introduced by contamination at some point in the preparation chain [5]. To discipline PAL, several organisms propose the adoption of a risk-based approach [6]. According to the Voluntary Incidental Trace Allergen Labelling (VITAL) system, a voluntary scheme developed by the Australian and New Zealand food industries [7], food industry may choose to renounce to precautionary statement (action level one) when a food allergen contamination is unlikely, or to include the statement ' . . . . may be present' (action level two) depending on the circumstances. The VITAL reference doses for specific food allergens are derived from diagnostic Oral Food Challenges (OFCs) in populations of patients with food allergy. For a biscuit that may contain egg and/or milk, action level one is suggested at a reference dose of 0.2 mg total protein. For a reasonable portion size of 50 g of biscuits, this translates in a suggested exemption from PAL when the concentration is of 4 mg total protein/kg or less [8,9].

As threshold doses have been not calibrated on a population of patients with severe food allergy, but on the entire population of food allergy sufferers, the risk thresholds may be over-evaluated [10]. Although only a part of the milk/egg allergic patients are also reactive to baked foods [11], the current thresholds were derived from OFCs without distinction among baked-tolerant and baked-allergic patients. Those allergic to baked forms are considered the most reactive portion of the milk/egg allergic population [12], but their thresholds have been only rarely compared to those of milk/egg allergic patients tolerant to baked foods [13].

In this scenario, studies on the effective clinical relevance of smaller doses than the VITAL thresholds in patients extremely allergic to milk and egg are lacking. For this reason, we wanted to evaluate in a group of baked egg- and/or baked milk-allergic patients the tolerance of a baked product labelled as 'may content little amounts of milk and egg'. A secondary objective of our study was to verify the protein quantity of milk and egg in the product using different analytical methods, in order to establish the relationship between quantities traceable below the 1% threshold and the possible development of symptoms.

Finally, this study offers us the opportunity to evaluate the threshold of reactivity to milk- and egg-baked proteins in children severely allergic to these foods.

2. Caseload and Methods

2.1. Patients

Between January 2016 and December 2019, pediatric patients aged 6 months–18 years affected by IgE-mediated milk and/or egg allergy were consecutively evaluated for their reactivity to baked milk/egg at Allergy Division of Bambino Gesù Children's Hospital in Rome. Patients with a history of immediate (<2 h) reactions to baked milk and/or egg sensitized to baked milk and/or egg using skin prick test (SPT) and with a positive specific IgE determination for milk, egg and/or their fractions, were included. For those without a recent convincing history of anaphylaxis, a confirmatory OFC with baked milk and/or egg was required. Children with unstable asthma and severe uncontrolled eczema were included when clinically stabilized.

2.2. Study Design

In a monocentric, prospective design, the patients were exposed to double-blind, placebo-controlled food challenge (DBPCFC) with milk-free, egg-containing or egg-free, milk-containing biscuits for confirmation of baked egg or baked milk allergy, respectively. The usual contraindications to OFCs were applied [14]: when an anaphylactic reaction had occurred <6 months before inclusion into the study for children of 0.5 to 5 years; <12 months for children aged 6–12 years, and <2 years for adolescents, the clinical history was considered sufficient proof of allergy to baked milk or egg. Such patients were not exposed to any confirmatory DBPCFC.

Children allergic to baked proteins were tested for their tolerance to egg-free, milk-free biscuits at OFC and SPT.

2.3. Diagnostic Challenges

DBPCFCs performed to confirm food allergy were calibrated up to 50 g of biscuit, corresponding to 4 g of baked milk or egg proteins. We used milk-free, egg-containing biscuits ("Pavesini", Barilla G. e R. Fratelli S.p.A., Parma, Italy; 2.09 g egg protein per 100 g product) for confirmation of baked egg allergy, and egg-free, milk-containing biscuits ("Biscottino Primi Mesi", Plasmon, Milano, Italy; 1.30 g milk protein per 100 g product) for confirmation of baked milk allergy. The challenges were administered in seven growing doses, with an initial dose of 0.25 g, corresponding to 5.20 mg egg or 3.25 mg milk protein, respectively. We proportioned the challenge doses to the reasonable dose for each age, using a lower dose in younger children and increasing the food amount up to 45.75 g in adolescents (Table S1). The foods were blinded according to the standardized AAAAI/Europrevall protocol [15,16]. The oral challenges were discontinued at the first onset of objective symptoms [17]. Patients were observed up to 6 h after starting the test.

2.4. Milk/egg IgE Sensitization

SPT with cow's milk, casein, egg white, egg yolk (Lofarma, Milano, Italy), fresh cow's milk, fresh egg white, Pavesini, and Biscottino Primi Mesi were performed. In this case, 10 mg/mL histamine phosphate in 50% glycosaline and glycosaline on its own (Lofarma, Milano, Italy) were used as positive and negative controls, respectively. A Dome-Hollister-Stier lancet with a 1 mm tip was used for the procedure. Wheal diameters were read through a clear plastic calliper disk scaled in mm under × 4 magnification, and were interpreted when the wheal margin was included within a complete caliper circle to the nearest mm [18]. A limit of 5 mm was set for SPT positivity.

Ten mL of venous blood was collected from the patients to determine serum IgE levels (total and specific for cow's milk proteins, casein, egg white, egg yolk) using ImmunoCAP® (Thermo Fisher Scientific, Waltham, MA, USA), with 0.35 kU/L as a lower limit of eligibility [19]. Part of the serum was stored at −80 °C for successive protein sensitivity evaluations.

2.5. Evaluation of the Clinical Tolerance to the Biscuits Labelled without Milk and Egg

Within one month of inclusion in the study, the selected patients underwent SPT and OFC with milk and egg –free biscuits "Magretti" [Galbusera S.p.A., Cosio Valtellino (SO), Italy] from 12 different lots. Such biscuits are labelled as "not containing milk and egg". However, the label indicates "it cannot be excluded that any traces of these allergens are present, in any case less than 5 mg/kg" (Table 1).

Based on the age of patients, seven or eight incremental doses of food were administered at 20-min intervals under clinical supervision. The first six doubling doses, starting from 0.25 g up to 8.00 g, were the same for all study participants. The seventh and eighth doses differed in different age groups (Table S1). A sample of the biscuits used for the OFC and SPT was stored for proteomic evaluation. A symptoms-based clinical score assessing the degree of gastro-intestinal, respiratory, cardiovascular and dermatological reactions was applied to monitor the acute allergic reactions. The procedure was interrupted when clear-cut objective symptoms were present, or after any severe, persistent (over 40 min) subjective symptom, according to our stopping rules [17].

Table 1. Magretti biscuit composition.

Ingredient	Quantity
Type 2 soft wheat flour	63%
Sugar	
Cereal flour	10% (corn 5%, barley 5%, on the finished product)
High oleic sunflower oil	10%
Honey	
Barley malt and corn extract	
Raising agents	ammonium acid carbonate, disodium diphosphate, sodium hydrogen carbonate
Whole sea salt	0.5%
Emulsifier	
Aromas	
Allergy warning	The recipe does not contain milk and eggs. It cannot be excluded that any traces of these allergens are present, in any case less than 5 mg/kg. The product can also contain soy, hazelnuts and other nuts; therefore, it is not suitable for consumption by people allergic to these substances.

2.6. Evaluation of the Presence of Milk/Egg Traces in Biscuits Labelled without Milk and Egg

Biscuits samples used in OFC and SPT were stored at $-80\ ^{\circ}$C in powder form and analyzed by immunoblot and Liquid Chromatography–ElectroSpray–tandem Mass Spectrometry (LC-ESI-MS/MS) for the detection of very small amounts of milk and egg allergens. The samples underwent the proteomic workflow for allergen detection and quantification as depicted in Figure 1. We adopted two different strategies in order to assess the contamination by egg and milk allergens:

1. immunoblot, to determine patient serum binding to egg and milk allergens possibly contained in tested biscuits (Section 2.6.1);
2. two different LC-ESI-MS/MS methods, aimed at quantifying egg and milk allergens in biscuits by monitoring their marker peptides (Section 2.6.2).

2.6.1. Protein Extraction, Gel Electrophoresis Separation and Immunoblot Analysis

Biscuits were first ground coarsely using a mortar and pestle, and then milled mechanically for 30 s (three times, 10 s) in a blender (IKA Werke GmbH, Staufen im Breisgau, Germany). The biscuits' powder was weighted and 1 g of sample was combined with 20 mL of extraction buffer (50 mM Sodium carbonate/bicarbonate pH 9.6), then it was left shaking for 2 h at 60 $^{\circ}$C. The extract was sonicated for 5 min, 4 s pulse and 4 s pause, 60% amplitude (VibraCell Ultrasonic Liquid Processor, Sonics and Materials Inc, Newtown, CT, USA) and centrifuged for 10 min at 3000 g at room temperature (r.t.). In this case, 10 mL of supernatant were filtered through first an Acrodisc 25 mm syringe filter by a 1.2 μm Versapor membrane (Pall Corporation, Ann Arbor, MI, USA) and then by a 0.45 μm acetate cellulose membranes (Minisart Syringe Filter, Sartorius Stedim Biotech GMbh, Goettingen, Germany). Four mL of the filtrate were loaded on centrifugation filter devices [Amicon Ultra, 3000 Da molecular weight cut-off (MWCO); Merck Millipore, Billerica, MA, USA)], and concentrated 10 times. Protein concentration of the filtered and concentrate sample was determined with the colorimetric Bicinchoninic Acid Assay (Thermo Fisher Scientific). One-dimensional sodium dodecyl sulfate-polyacrylamide gel electrophoresis (SDS-PAGE) of samples was performed loading 40 μg of extracted proteins onto 12% Bis-Tris Criterion XT precast gels (11 cm, Bio-Rad Laboratories, Hercules, CA, USA); separation was performed in running buffer (25 mM Tris, 192 mM Glycine, 0.1% SDS) on a Criterion Cell apparatus (Bio-Rad) at 120 V.

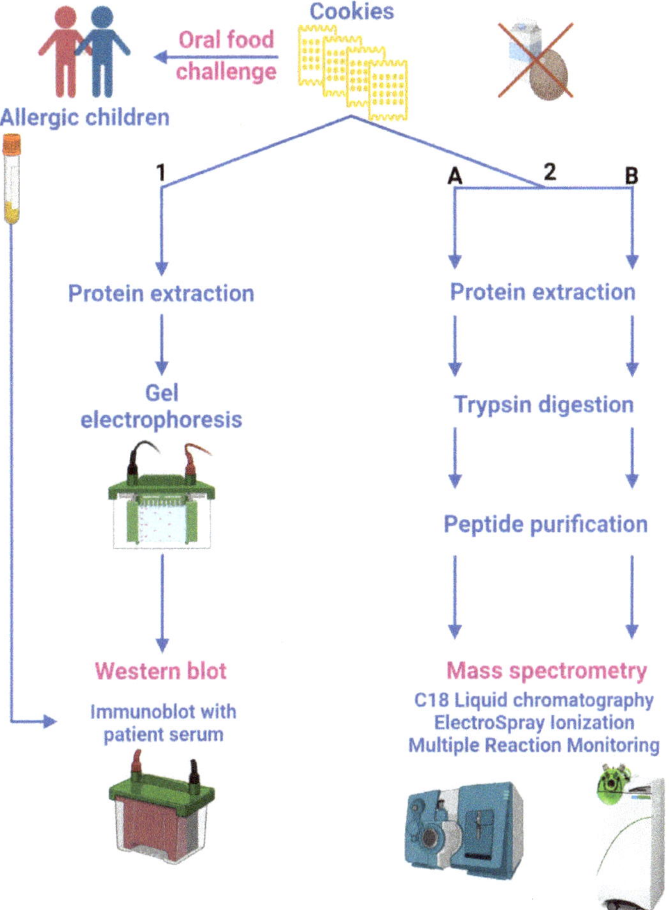

Figure 1. Sketch of the proteomic experimental approaches. (1) Immunoblot experiments determine patient serum binding to egg and milk allergens; (2) LC-ESI-MS/MS analyses, following two different workflows and different Triple Quadrupole MS platforms (2.A and 2.B), quantify egg and milk allergens in biscuits.

Immunoblotting was performed by transferring proteins from gel to polyvinylidene fluoride membrane (Bio-Rad) at 350 mA for 2 h, in a cold room at 4 °C, on Criterion™ Blotter with wire electrodes (Bio-Rad) in presence of transfer buffer (25 mM Tris, 192 mM Glycine, 20% Methanol). The membranes were left in contact with a solution of Pierce™ Protein-Free T20 Blocking Buffer (Thermo Fisher Scientific) and then incubated overnight at 4 °C in serum of each study subject, diluted 1:10 in Blocking Buffer. After several rinses with phosphate buffer (200 mg/L KCl, 200 mg/L KH_2PO_4, 8000 mg/L NaCl, 1150 mg/L Na_2HPO_4)-0.1%Tween), the membrane was incubated for two hours at r.t. with a secondary anti-human IgE antibody conjugated to the enzyme alkaline phosphatase (Mouse Anti-Human IgE Fc-AP, clone B3102E8; Southern Biotech, Birmingham, AL, USA) diluted 1:500 in Blocking Buffer. The membrane was washed again in PBS-T and the immunoreactive signals were detected by a colorimetric reaction in presence of 5-bromo-4-chloroindolyl phosphate/Nitroblue Tetrazolium and 0.1 M Tris, 0.5 mM MgCl2 (pH 9.5) (Alkaline Phosphate Conjugate Substrate Kit, Bio-Rad). Stained membranes were scanned with a ChemiDocTM XRS⁺ Molecular Imager (Bio-Rad). SDS-PAGE protein molecular weight standards, to monitor the run, and proteins from biscuits containing egg and milk ingredients ("BuoniCosì", Galbusera), as positive control, were loaded. Analyses were performed on two different portions of Magretti biscuits previously administered to enrolled patients. (Table S2). Unspecific signals due to the secondary antibody were identified performing immunoblots without patients' sera or with pooled sera of non-allergic pediatric patients. A Relative Volume Quantity analysis was performed by a densitometric analysis of blots using Image Lab Software (version 6, Bio-rad). Lane of positive control (biscuits containing egg and milk) was set as reference, and for all other lanes of the blot numerical values relative to the reference were calculated as ratio of the background-adjusted lane volume divided by the background-adjusted reference volume.

2.6.2. LC-ESI-MS/MS Analysis

Two different procedures were followed, route 2.A or 2.B (Figure 1), in order to apply different workflows and instrumental platforms with the aim to validate the final results.

Method 2.A: One g of biscuits' powder was combined with 5 mL of Hexane, left shaking at high speed for 15 min at r.t., centrifuged at 4000 g for 20 min, and the supernatant discarded. This lipid removal procedure was repeated once again and samples were dried by evaporation using a flow of nitrogen. Four mL of extraction buffer [50 mM Trizma base, 2 M Urea, 1% (*w/v*) Octyl ß-D-glucopyranoside] were added to each defatted sample, which was shaken at high speed for 1 h at r.t. After centrifugation, 500 µL of supernatant (protein extracts) were reduced for 60 min, shaking at 1000 rpm at 60 °C with 5 mM (final concentration) Tris-(2-carboxyethyl)-phospine, incubated with 25 µL of 100 mM cysteine blocking reagent (Methyl methane-thiosulfonate) for 15 min at r.t., diluted with 425 µL of digestion buffer (100 mM Ammonium bicarbonate, 5 mM Calcium chloride), and digested with 20 µg Trypsin TPCK treated (SCIEX, Redwood City, CA, USA) at 37 °C for 16 h. The reaction was stopped with 30 µL of Formic acid (FA) and the digested samples were filtrated using a centrifugal filter unit with 10 kDa MWCO (Merck Millipore). The filtrates were further analysed by mass spectrometry. In order to obtain a calibration curve spanning the concentration range 5–50 part per million (ppm, defined as µg of allergenic ingredient per g of matrix) for quantitative analysis, biscuits fortified with allergen commodity (egg powder and skim milk powder, certified reference material BCR-685, Sigma-Aldrich, Milan, Italy) were prepared and processed as samples.

Digested samples were analyzed using a micro-LC-ESI-Triple Quadrupole (TQ) platform: a M3 MicroLC-TE System interfaced with a QTrap6500⁺ mass spectrometer equipped with an IonDrive Turbo V Ion Source (TurboIonSpray probe with 50 µm i.d. electrode; Sciex). In this case, 10 µL of tryptic peptides (~40 µg) were injected onto a ChromXP C18 trap column cartridge (5 µm × 10 mm, 120 Å, 300 µm i.d.; Sciex) for pre-concentration and desalting at a flow rate of 50 µL/min, and subsequently separated using a HALO peptide-ES C18 column (300 µm × 150 mm, 160 Å, 2.7 µm; Advanced Materials Technology, Wilmington, DE, USA) maintained at 40 °C. Mobile phase A was H_2O + 0.1% FA, and mobile phase B was 0.1% FA in acetonitrile (ACN). Peptides were separated by linear gradient of 2–40% mobile phase B over 11 min at a flow rate of 10 µL/min, followed by a 2 min rinse with 98% mobile phase B. The column was re-equilibrated at the initial conditions for 5.4 min. The QTrap mass spectrometer was operating in positive ESI High Masses Multiple Reaction Monitoring (MRM) mode (Unit Resolution on Q1 and Q2); the data were acquired using Analyst (version 1.6.3, Sciex). Source/Gas parameters were: 20 psi Curtain Gas; Medium Collision Gas; 5500 V IonSpray Voltage; 150 °C Temperature; 35 psi Ion Source Gas 1; 20 psi Ion Source Gas 2, and instrumental settings optimized for each individual milk and egg peptide marker are reported in Table S3. Analyses were performed on two different portions of Magretti biscuits used for the OFC and SPT (Table S2).

Method 2.B: A subset of samples (from cookies administered to child patients namely n. 8, 9, 11, 13–14, 16–21, 23, 25–28) underwent a different sample preparation workflow and a MRM method was built up on a different platform: an LX50 UHPLC pump provided with an autosampler and an ESI interface connected to a QSight 220 TQ mass spectrometer (PerkinElmer Inc., Waltham, MA, USA) as already published [20].

Instrumental settings optimized for each individual milk and egg peptide marker are reported in Table S4. All analyses were performed in triplicates.

Allergen-free and incurred biscuits at the highest level of 300 µg allergenic ingredient/g matrix were produced at laboratory scale according to the procedure described elsewhere [21]. Lower concentration levels covering the calibration range as previously indicated in route 2.A were produced from these two stock samples by appropriate dilution with blank matrix powder. The allergen-free biscuit digest was fortified with increasing amount of synthetic standard peptides (GenScript, Piscataway, NJ, USA) specific for milk and egg allergens (in the range 0.0125–0.2500 µg/mL) and calibration curves were prepared by plotting the signal of each candidate peptide against the inclusion level in the biscuit sample. All extracts were submitted to the workflow previously described before its injection (10 µL) in duplicate in the QSight equipment.

For quantification purposes, each synthetic peptide with peptide concentration (expressed as µg/mL) was first converted in protein molarity, assuming that full digestion of the protein took place and then a proper conversion factor was applied for the calculation taking into consideration the mass/volume ratio used for protein extraction.

Quantitative analysis on data obtained by QTrap mass spectrometer was performed by MultiQuant software (version 3.0.2, Sciex) applying MQ4 algorithm for peak integration (minimum Gaussian smooth width of 1 point) and data processing. Calibration curves were generated by plotting peak areas against allergen commodity concentrations, with $1/x$ fitting. In particular, calibration points were produced spanning one order of magnitude concentration range expressed as µg allergenic ingredient/g matrix.

Peak integration and data processing on QSight 220 spectrometer MS data was performed by using 3Q Simplicity software (version 1.4, Perkin Elmer) applying Moving Average algorithm for peak integration (minimum Gaussian smoothing of 5 point).

The reporting units were converted into total proteins of allergenic ingredient (µg/g) assuming 35.39% and 48.00% of total protein content for milk and egg, respectively, in accordance with what reported by USDA.

2.7. Statistics

As our objective was to verify whether Magretti may be considered technically hypoallergenic for a population of children allergic to baked egg or milk, our sample size was calculated according to the American Academy of Pediatrics (AAP) guidelines for clinical testing of hypoallergenic formulas [22]. The number of subjects needed to project with 95% confidence (one-sided interval) that less than 10% of infants will react to the product is 29 if no clinical reactions are observed, and 43 if one clinical reaction is observed. These sample size estimates were derived based on binomial distribution techniques using Wald's method for deriving confidence intervals for single proportions.

The analysis on the primary outcome parameter was a per protocol (PP) analysis. Additional outcomes were obtained on Full Analysis Set (FAS) based on the intention-to-treat (ITT) assumption. Quantitative parameters have been summarized by descriptive statistics (arithmetic mean, standard deviation, minimum, median, and maximum) and qualitative parameters by frequencies and percentages. Categorical variables have been presented using non-missing observations and percentages. Denominators for calculation of percentages have been taken as the number of subjects with non-missing observations in the specified population unless otherwise stated. Continuous variables have been presented using number of subjects in the analysis population (N), number of subjects with non-missing observations (n), mean, standard deviation (abbreviated as "SD" in statistical tables), median, minimum (Min) and maximum (max). Unless stated otherwise, statistical tests have been conducted as two-sided at a level of significance $p = 0.05$. p-values for difference from baseline have been calculated using paired t-test.

Statistical analyses were performed using the SPSS statistical software, version 19.0 (SPSS Inc., Chicago, IL, USA).

3. Results

3.1. Clinical Characteristics

Over the four-year period considered, among 379 children confirmed with milk/egg allergy at our hospital, 152 patients were reactive with baked milk/egg. In this case, 41 of them met the severity criteria. Nine patients were excluded: two did not accept the confirmatory challenge, six returned negative at DBPCFC, and one had celiac disease.

The test was proposed to 32 children, positive at entry OFC with Plasmon (23) and/or Pavesini (12). Two families did not agree to participate in the study, and one withdrew her consent on the day of the test when the child refused to perform the blood test. Hence, 29 children (17 male and 12 female, median age 6.97 years, range 0.67–16.75 years) were included in the study population (Table S5). Their clinical characteristics are reported in Table S6. In this case, 25 were allergic to milk (21 had positive SPT to milk-containing biscuits), 19 to egg (11 SPT-positive to egg-containing biscuits); 15 patients were allergic to both foods, three of whom returned positive to both baked egg and milk (Table 2).

Table 2. Patients' sensitization to milk and egg (number of patients, #; mean $\pm$ SD of sensitization values).

	sIgE > 0.35 kUI/L		c-SPT > 3 mm		ffSPT > 3 mm	
	#		#		#	
Milk	23	48.4 ± 39.7	24	10.3 ± 5.3	25	11.4 ± 4.5
Casein	19	49.1 ± 45.4	15	9.7 ± 4.9		
Egg white	15	27.1 ± 46.2	19	7.6 ± 2.5	15	7.0 ± 2.2
Egg yolk	9	32.4 ± 45.2	14	9.0 ± 3.5	10	8.0 ± 1.2
Baked milk biscuit (Plasmon)					21	7.2 ± 3.8
Baked egg biscuit (Pavesini)					11	4.7 ± 1.2
Magretti-Frollini con orzo e mais biscuit					0	

SD: standard deviation; sIgE: specific IgE; c-SPT: commercial skin prick test; ffSPT: fresh-food skin prick test; n.d.: not determined.

Five out of the 29 included patients, were not exposed to confirmatory food challenges due to recent anaphylaxis (three in the group under two years, one in the group 5–13 years and one in the group 13–18 years). The remaining 24 underwent DBPCFCs with baked milk (15 cases), baked egg (five cases), or both (four cases). From these challenges, the mean thresholds of reactivity were 116.3 ($\pm$ 107.6) and 128.3 ($\pm$ 96.7) mg protein for milk and egg, respectively.

No patient presented any symptoms at any challenge time during OFCs with the tested product. Equally negative were all skin tests: no patient resulted SPT-positive to Magretti.

3.2. Determination of Patient Serum Binding to Egg and Milk Allergens Contained in Biscuits

The mean total soluble protein recovery was 7.7 mg per gram. Immunoblots did not show reactions between serum of 18 patients and proteins contained in biscuits (Figure 2). In biscuits from patients 3, 6, 12, 15, 18, 20, 22, 23, 25, 26 and 29, non-specific signals were detected accounting for 30.12% of the positive control.

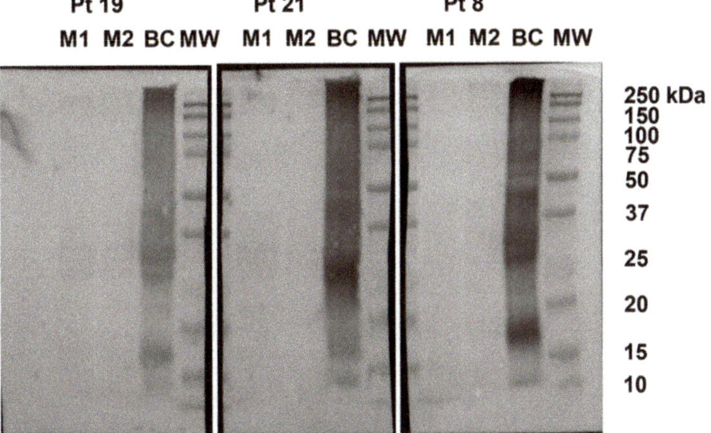

Figure 2. Representative immunoblots. Proteins extracted from two doses (M1 and M2) of biscuit tested on patients (Pt) 19, 21, and 8, blotted with the corresponding patient serum. Proteins extracted from biscuits containing egg and milk ingredients (BC) were loaded as positive control. MW: protein molecular weight standards.

3.3. Determination of Cow's Milk and Hen's Egg Allergen Levels in Commercial "Milk and Egg Free" Biscuits by Targeted Mass Spectrometry Methods

Using method 2.A, a mean of ~10 mg of protein was extracted from 1 g of biscuits. About 4 mg of each protein extract was subjected to reduction, alkylation and protein digestion. In method 2.B a different sample preparation workflow was used. In both methods, peptides from milk αS1-casein and egg ovalbumin allergens were used as marker proteins by MRM analysis.

To evaluate the sensitivity of the method, matrix-matched calibration curves, obtained by fortifying biscuits with increasing amounts of allergenic ingredients to cover the range 5–50 mg of allergens/kg of matrix, were built up for each milk and egg allergen marker selected. By interpolating the data of the matrix matched calibration curve, the linearity over the concentration range investigated for all the markers monitored was verified, with a correlation coefficient always better than 0.98. Finally, limit of detection (LOD) and quantification (LOQ) values were, respectively, calculated as 3 and 10 times the standard deviation of the line intercept divided by the slope of the calibration equation for both methods.

Table 3 reports the calculated LODs for methods 2.A and 2.B depending on the instrumental platform used and the selected transitions monitored. Using method 2.A, we were able to detect traces of milk and egg allergens at the lowest range of 0.6 µg $_{tot\,prot}$/g $_{matrix}$ for milk and egg. Method 2.B was able to quantify milk and egg allergen in biscuits at the lowest level 0.1 and of 0.3 µg $_{tot\,prot}$/g $_{matrix}$, respectively.

Once the methods had been optimized, they were applied to Magretti to verify their allergen contamination at the lowest level offered by the method. The two methods showed a good agreement of the results obtained. None of the analyzed samples was found contaminated with milk and egg according to the sensitivity offered by the MS method (Table 3).

Table 3. Results of MRM experiments referred to matrix-matched calibration curves produced in fortified biscuits (route 2.A and route 2B) employing synthetic peptides for quantification (in route 2.B).

Allergen	Protein	Quantifier Peptide (*m/z*)	Product ion (*m/z*)	LOD μg tot protein/g matrix	R^2	Route
Milk	α-S1-Casein Bos d9	634.4 (YLG)	991.6	0.63	0.99	2.A (QTrap 6500$^+$)
		692.9 (FFV)	991.4	0.10	0.99	2.B (QSight 220)
Egg	Ovalbumin Gal d2	844.4 (GGL)	666.3	0.61	0.98	2.A (QTrap 6500$^+$)
		592.1 (ISQ)	858.9	0.30	0.99	2.B (QSight 220)

m/z: mass-to-charge ratio of the peptide ion and product; LOD: limit of detection calculated as $3 \times$ SD of the intercept calculated on the matrix matched calibration curve and whose goodness of the linear interpolation is reported by R2.

4. Discussion

Allergy to baked milk/egg occurs in a minority of patients allergic to the respective native foods. In previous experiences, this proportion is around 30% [23–26]. On the total allergic patients enrolled in this study 40.1% of children with milk/egg allergy were reactive to baked foods. This higher percentage may reflect a high severity of our caseload, afferent to a third-level hospital with a catchment area corresponding to the entire Italian nation. To focus on the most severe forms of baked egg/milk allergy, we applied an arbitrary definition of severity including clinical data and sensitization parameters. Basing on this, one fourth was defined severe. Most of the severe milk/egg allergic patients (21/29) had a history of anaphylaxis. In these highly selected patients, we found thresholds of 116.3 and 128.3 mg protein for milk and egg, respectively, at DBPCFC. As in previous studies these thresholds ranged between 0.6 and 150 mg for milk and between 0.65 and 200 mg for egg [27], our patients severely allergic to baked foods do not present thresholds below those allergic to milk/egg, confirming previous data [13,28,29].

In the studied population, no signs or symptoms of allergic reactions were recorded at OFC with the milk/egg free biscuits. In the light of quantitative assessments, appears that Magretti does not pose a real danger from possible accidental contamination.

In this model, can we predict the absence of risk simply using an accurate quantification of milk/egg allergens contained in foods? Probably we can.

According its labelling, Magretti may contain up to 5 mg/kg milk/egg protein, an amount exceeding the VITAL threshold by 20%. In principle, they are likely to induce allergy in up to 1% of milk/egg allergic individuals, and more in patients with severe food allergy. By analysing biscuits by LC-ESI-MS/MS methods, no sample was found contaminated at levels close to the 5 mg/kg indicated by the producer, corresponding, respectively, to 1.77 μg milk or 2.40 μg egg proteins/g of matrix. In addition, the VITAL 3.0 reference dose for milk and egg was not even remotely approached. We found a milk and egg protein content < 0.6 μg /g matrix of milk (method A), <0.1 μg /g matrix of milk and <0.3 μg/g matrix of egg protein (method B). As we administered between 31.75 and 45.75 g Magretti biscuits during OFCs, our patients were exposed to a maximum of 3.18/4.58 μg milk and a maximum of 9.53/13.64 μg egg protein, respectively. These values are much lower than the VITAL 3.0 limit of 200 micrograms for the same amount of milk or egg, under which precautionary cross-contact statement is not required. We can therefore confidently assume that mass spectrometry is able to confirm the absence of protein allergens up to a calculated level thus assuring a high level of safety for our patients. If this were the case, biscuit producers could be advised not to adopt any PAL for products containing such tiny amounts of milk or egg proteins: the risk would be far lower than the 1% predicted by VITAL grids.

Among omics sciences, proteomics and particularly MS-based proteomics are gaining a steadily increasing interest by the scientific community thanks to the recent and rapid technological advances made: up-to-date mass spectrometers have risen unprecedented specificity, sensitivity and capability to perform multiplexing analysis for allergen determination through their peptide/protein markers. For food control, MS-based proteomics approaches are currently applied for the detection and quantification of allergenic ingredients intended ads contaminants. In this regard, mass spectrometrists are making great efforts to develop allergens accurate quantification methods; MS strength lies in its ability to unequivocal identify allergens and multiplex the analyses, opening to the quantification of several allergenic proteins in complex matrices within a single LC-ESI-MS/MS run with high analytical confidence [30,31].

Aware of the technical difficulty of the proteomic methods, and of the likely bias among the different laboratories, we designed this study including analysis carried out in two analytical laboratories using different analytical platforms in order to compare the results originated by different analytical strategies and monitoring specific peptides/transitions for each selected allergen. On this regard, the two applied proteomic methods (2.A and 2.B, Figure 1), based on Triple Quadrupole mass spectrometry detection, followed a different analytical workflow and used different peptide transitions as quantifier ions, but did not provide significantly different values (Table 3). We infer that the standardization of proteomic methods may allow the analytic window necessary for an almost complete exclusion of allergic risk. By contrast, immunoblotting is in our hands too coarse to be able to contribute to the necessary information in this field, because it is burdened with interference errors.

Ideally, this study should have been conducted on 'very small amounts of egg/milk reactors'. As our first challenge dose was of 3.5 mg of milk protein or 5.2 mg of egg protein, we are not including patients positively allergic to very small amounts. Reaction to traces of milk and egg is an exceptional phenomenon, happening in less than 1% of food-allergic patients by VITAL definition. In order to be able to transfer the same study to a population of trace-allergic patients, it would be necessary to have a basic series of 2900 patients allergic to baked milk or egg, which is not available to us and largely exceeds any caseload ever published.

A second limitation is that we were not able to stratify patients based on a shared definition of food allergy severity. In the present situation, a precise classification of the different phenotypes of food allergy in a homogeneous way between different caseloads is not possible. The imminent definition of severe food allergy could help fill this unmet need [32–34].

A third limitation of our study could be the use of open OFCs in the evaluation of reactivity to the tested biscuit. The results deriving from this type of OFC can be different those of DBPCFC in diagnostic terms. However, it has been shown that they can be overlapping in terms of evaluation of food allergen dose distribution [35].

A further limitation of this study is that the MS-based approaches we used for the detection and quantification of allergenic ingredients are able to detect an amino acid sequence of the allergenic protein, but not necessarily the epitope recognized by immune system. Thus, it may be theoretically possible that they miss small parts of allergenic proteins. On the other hand, the presence of a peptide marker will definitely imply the presence of a milk or egg allergen traces giving rise to assume that a cross-contact with the allergen sources has occurred. According to its well-known selectivity and sensitivity, MS have the power to overcome immunoassays (enzyme-linked immunosorbent assay, ELISA) and PCR-based techniques, the historically most adopted methods for allergens detection and quantitation. Specifically, although ELISA methods boost a general high sensitivity, they still encounter disadvantages such as cross-reactions with food matrix, limited reproducibility, variable specificity of antibodies in the commercial kits, lack of standard reference materials for some allergens and missing multiplexing detection ability [36].

5. Conclusions

Due to the technological limitations, the current approach to PAL relies on a non-analytic-based risk assessment. As in our study the sensitivity of MS proteomic largely exceeding the limits recommended by the VITAL grid, we conclude that an accurate quantification of tiny amounts of protein in complex foods, in combination with population clinical studies, deserves the potential to establish exact reference doses below which no reactions could be exerted even in the most sensitive individuals.

When proteomic determinations show that the controls carried out at the level of the production and distribution chain are sufficient to avoid this risk for the tested product, the clinician may be authorized to exempt children allergic to milk and egg from observing PAL.

Supplementary Materials: The following are available online at https://www.mdpi.com/article/10.3390/nu13124540/s1, Table S1: Doses of biscuit administered to children in the experimental challenge, Table S2: Correspondence amongst Magretti doses tested on patients by OFC and SPT and samples subjected to proteomic experiments, Table S3: QTrap 6500+ MS compound parameters of milk α-S1-Casein and egg albumin detected ions, Table S4: QSight 220 MS compound parameters of milk α-S1-Casein and egg albumin detected ions, Table S5: Demographic data of the patients, Table S6: Clinical characteristics of the patients.

Author Contributions: Conceptualization, V.F., A.F. and L.P.; methodology, V.F., V.M., L.M., R.P., A.F. and L.P.; validation, L.M., A.F. and L.P.; formal analysis, R.V.; investigation, V.F., V.M., L.M., P.V., M.M., S.L.M., R.P., L.D., V.C., G.B. and C.R.; resources, L.M., A.F. and L.P.; writing—original draft preparation, V.F., M.M. and A.F.; writing—review and editing, V.M., L.M., R.P. and S.A.; visualization, V.M.; supervision, A.F and L.P.; project administration, A.F.; funding acquisition, A.F. and L.P. All authors have read and agreed to the published version of the manuscript.

Funding: This research was funded by Galbusera S.p.A. It also was supported by a grant of Italian Ministry of Health, Ricerca Corrente 2018, to Lorenza Putignani and Alessandro Fiocchi.

Institutional Review Board Statement: The study was conducted according to the guidelines of the Declaration of Helsinki, and approved by the Institutional Ethics Committee of Bambino Gesù Children's Hospital IRCCS (protocol number 1022_2015_OPBG, 10 December 2015).

Informed Consent Statement: Written informed consent from either parents or legal representative of children was obtained for the participation to the study.

Data Availability Statement: Dataset is available on request.

Acknowledgments: We would like to thank P. Dossetto and L. Rieux (Sciex) for helpful advice.

Conflicts of Interest: The authors declare no conflict of interest. The funders had no role in the design of the study; in the collection, analyses, or interpretation of data; in the writing of the manuscript, or in the decision to publish the results.

References

1. Roberts, G.; Allen, K.; Ballmer-Weber, B.; Clark, A.; Crevel, R.; Galvin, A.D.; Fernandez-Rivas, M.; Grimshaw, K.E.C.; Hourihane, J.O.; Poulsen, L.K.; et al. Identifying and managing patients at risk of severe allergic reactions to food: Report from two iFAAM workshops. *Clin. Exp. Allergy* **2019**, *49*, 1558–1566. [CrossRef] [PubMed]
2. Available online: https://Farrp.Unl.Edu/Documents/Regulatory/International_%20Allergens_061021.Pdf (accessed on 4 November 2021).
3. Fierro, V.; Di Girolamo, F.; Marzano, V.; Dahdah, L.; Mennini, M. Food labeling issues in patients with severe food allergies: Solving a hamlet-like doubt. *Curr. Opin. Allergy Clin. Immunol.* **2017**, *17*, 204–211. [CrossRef]
4. Allen, K.J.; Turner, P.J.; Pawankar, R.; Taylor, S.; Sicherer, S.; Lack, G.; Rosario, N.; Ebisawa, M.; Wong, G.; Mills, E.; et al. Precautionary labelling of foods for allergen content: Are we ready for a global framework? *World Allergy Organ. J.* **2014**, *7*, 1–10. [CrossRef] [PubMed]
5. Ford, L.S.; Taylor, S.L.; Pacenza, R.; Niemann, L.M.; Lambrecht, D.M.; Sicherer, S.H. Food allergen advisory labeling and product contamination with egg, milk, and peanut. *J. Allergy Clin. Immunol.* **2010**, *126*, 384–385. [CrossRef] [PubMed]
6. National Academies of Sciences, Engineering, and Medicine; Health and Medicine Division; Food and Nutrition Board; Committee on Food Allergies. *Global Burden, Causes, Treatment, Prevention, and Public Policy Finding a Path to Safety in Food Allergy: Assessment of the Global Burden, Causes, Prevention, Management, and Public Policy*; Oria, M.P., Stallings, V.A., Eds.; National Academies Press: Washington, DC, USA, 2016.

7. Allen, K.J.; Remington, B.; Baumert, J.L.; Crevel, R.W.; Houben, G.; Brooke-Taylor, S.; Kruizinga, A.G.; Taylor, S.L. Allergen reference doses for precautionary labeling (VITAL 2.0): Clinical implications. *J. Allergy Clin. Immunol.* **2014**, *133*, 156–164. [CrossRef]
8. Remington, B.C.; Westerhout, J.; Meima, M.Y.; Blom, W.M.; Kruizinga, A.G.; Wheeler, M.W.; Taylor, S.L.; Houben, G.; Baumert, J.L. Updated population minimal eliciting dose distributions for use in risk assessment of 14 priority food allergens. *Food Chem. Toxicol.* **2020**, *139*, 111259. [CrossRef]
9. Houben, G.F.; Baumert, J.L.; Blom, W.M.; Kruizinga, A.G.; Meima, M.Y.; Remington, B.C.; Wheeler, M.W.; Westerhout, J.; Taylor, S.L. Full range of population Eliciting Dose values for 14 priority allergenic foods and recommendations for use in risk characterization. *Food Chem. Toxicol.* **2020**, *146*, 111831. [CrossRef]
10. Madsen, C.B.; Dungen, M.W.V.D.; Cochrane, S.; Houben, G.F.; Knibb, R.; Knulst, A.C.; Ronsmans, S.; Yarham, R.A.; Schnadt, S.; Turner, P.; et al. Can we define a level of protection for allergic consumers that everyone can accept? *Regul. Toxicol. Pharmacol.* **2020**, *117*, 104751. [CrossRef]
11. Gruzelle, V.; Juchet, A.; Martin-Blondel, A.; Michelet, M.; Chabbert-Broue, A.; Didier, A. Benefits of baked milk oral immunotherapy in French children with cow's milk allergy. *Pediatr. Allergy Immunol.* **2020**, *31*, 364–370. [CrossRef]
12. Lemon-Mulé, H.; Sampson, H.A.; Sicherer, S.H.; Shreffler, W.G.; Noone, S.; Nowak-Wegrzyn, A. Immunologic changes in children with egg allergy ingesting extensively heated egg. *J. Allergy Clin. Immunol.* **2008**, *122*, 977–983.e1. [CrossRef]
13. Remington, B.C.; Westerhout, J.; Campbell, D.E.; Turner, P.J. Minimal impact of extensive heating of hen's egg and cow's milk in a food matrix on threshold dose-distribution curves. *Allergy* **2017**, *72*, 1816–1819. [CrossRef] [PubMed]
14. Grabenhenrich, L.B.; Reich, A.; Bellach, J.; Trendelenburg, V.; Sprikkelman, A.B.; Roberts, G.; Grimshaw, K.E.C.; Sigurdardottir, S.; Kowalski, M.L.; Papadopoulos, N.G.; et al. A new framework for the documentation and interpretation of oral food challenges in population-based and clinical research. *Allergy* **2016**, *72*, 453–461. [CrossRef]
15. Nowak-Wegrzyn, A.; Assa'Ad, A.H.; Bahna, S.L.; Bock, S.A.; Sicherer, S.H.; Teuber, S.S. Work Group report: Oral food challenge testing. *J. Allergy Clin. Immunol.* **2009**, *123*, S365–S383. [CrossRef] [PubMed]
16. Keil, T.; McBride, D.; Grimshaw, K.; Niggemann, B.; Xepapadaki, P.; Zannikos, K.; Sigurdardottir, S.T.; Clausen, M.; Reche, M.; Pascual, C.; et al. The multinational birth cohort of EuroPrevall: Background, aims and methods. *Allergy* **2010**, *65*, 482–490. [CrossRef]
17. Grabenhenrich, L.B.; Reich, A.; McBride, D.; Sprikkelman, A.; Roberts, G.; Grimshaw, K.E.C.; Fiocchi, A.G.; Saxoni-Papageorgiou, P.; Papadopoulos, N.G.; Fiandor, A.; et al. Physician's appraisal vs documented signs and symptoms in the interpretation of food challenge tests: The EuroPrevall birth cohort. *Pediatr. Allergy Immunol.* **2017**, *29*, 58–65. [CrossRef]
18. Fiocchi, A.; Bouygue, G.R.; Restani, P.; Bonvini, G.; Startari, R.; Terracciano, L. Accuracy of skin prick tests in IgE-mediated adverse reactions to bovine proteins. *Ann. Allergy Asthma Immunol.* **2002**, *89*, 26–32. [CrossRef]
19. Fiocchi, A.; Artesani, M.C.; Riccardi, C.; Mennini, M.; Pecora, V.; Fierro, V.; Calandrelli, V.; Dahdah, L.; Valluzzi, R.L. Impact of Omalizumab on Food Allergy in Patients Treated for Asthma: A Real-Life Study. *J. Allergy Clin. Immunol. Pract.* **2019**, *7*, 1901–1909.e5. [CrossRef]
20. Monaci, L.; De Angelis, E.; Guagnano, R.; Ganci, A.; Garaguso, I.; Fiocchi, A.; Pilolli, R. Validation of a MS Based Proteomics Method for Milk and Egg Quantification in Cookies at the Lowest VITAL Levels: An Alternative to the Use of Precautionary Labeling. *Foods* **2020**, *9*, 1489. [CrossRef]
21. Pilolli, R.; De Angelis, E.; Monaci, L. Streamlining the analytical workflow for multiplex MS/MS allergen detection in processed foods. *Food Chem.* **2017**, *221*, 1747–1753. [CrossRef] [PubMed]
22. American Academy of Pediatrics. Committee on Nutrition Hypoallergenic Infant Formulas. *Pediatrics* **2000**, *106*, 346–349. [CrossRef]
23. Agyemang, A.; Feuille, E.; Tang, J.; Steinwandtner, I.; Sampson, H.; Nowak-Węgrzyn, A. Outcomes of 84 consecutive open food challenges to extensively heated (baked) milk in the allergy office. *J. Allergy Clin. Immunol. Pr.* **2018**, *6*, 653–655.e2. [CrossRef]
24. Mehr, S.; Turner, P.J.; Joshi, P.; Wong, M.; Campbell, D.E. Safety and clinical predictors of reacting to extensively heated cow's milk challenge in cow's milk-allergic children. *Ann. Allergy Asthma Immunol.* **2014**, *113*, 425–429. [CrossRef]
25. Bartnikas, L.M.; Sheehan, W.J.; Hoffman, E.B.; Permaul, P.; Dioun, A.F.; Friedlander, J.; Baxi, S.N.; Schneider, L.C.; Phipatanakul, W. Predicting food challenge outcomes for baked milk: Role of specific IgE and skin prick testing. *Ann. Allergy Asthma Immunol.* **2012**, *109*, 309–313.e1. [CrossRef] [PubMed]
26. Nowak-Wegrzyn, A.; Bloom, K.A.; Sicherer, S.H.; Shreffler, W.G.; Noone, S.; Wanich, N.; Sampson, H.A. Tolerance to extensively heated milk in children with cow's milk allergy. *J. Allergy Clin. Immunol.* **2008**, *122*, 342–347.e2. [CrossRef]
27. Taylor, S.L.; Hefle, S.L.; Bindslev-Jensen, C.; Bock, S.; Burks, A.; Christie, L.; Hill, D.J.; Host, A.; Hourihane, J.O.; Lack, G.; et al. Factors affecting the determination of threshold doses for allergenic foods: How much is too much? *J. Allergy Clin. Immunol.* **2002**, *109*, 24–30. [CrossRef]
28. Dahdah, L.; Ceccarelli, S.; Amendola, S.; Campagnano, P.; Cancrini, C.; Mazzina, O.; Fiocchi, A. IgE Immunoadsorption Knocks Down the Risk of Food-Related Anaphylaxis. *Pediatrics* **2015**, *136*, e1617–e1620. [CrossRef] [PubMed]
29. Dantzer, J.A.; Dunlop, J.H.; Wood, R.A. Standard testing fails to identify patients who tolerate baked milk. *J. Allergy Clin. Immunol.* **2020**, *146*, 1434–1437.e2. [CrossRef]
30. Putignani, L.; Dallapiccola, B. Foodomics as part of the host-microbiota-exposome interplay. *J. Proteom.* **2016**, *147*, 3–20. [CrossRef] [PubMed]

31. Marzano, V.; Tilocca, B.; Fiocchi, A.G.; Vernocchi, P.; Mortera, S.L.; Urbani, A.; Roncada, P.; Putignani, L. Perusal of food allergens analysis by mass spectrometry-based proteomics. *J. Proteom.* **2020**, *215*, 103636. [CrossRef]
32. Fiocchi, A.; Ebisawa, M. Severe food allergies: Can they be considered rare diseases? *Curr. Opin. Allergy Clin. Immunol.* **2017**, *17*, 201–203. [CrossRef]
33. Arasi, S.; Nurmatov, U.; Turner, P.J.; Ansotegui, I.J.; Daher, S.; Dunn-Galvin, A.; Ebisawa, M.; Eigenmann, P.; Fernandez-Rivas, M.; Gupta, R.; et al. Consensus on DEfinition of Food Allergy SEverity (DEFASE): Protocol for a systematic review. *World Allergy Organ. J.* **2020**, *13*, 100493. [CrossRef] [PubMed]
34. Arasi, S.; Nurmatov, U.; Dunn-Galvin, A.; Daher, S.; Roberts, G.; Turner, P.J.; Shinder, S.B.; Gupta, R.; Eigenmann, P.; Nowak-Wegrzyn, A.; et al. Consensus on DEfinition of Food Allergy SEverity (DEFASE) an integrated mixed methods systematic review. *World Allergy Organ. J.* **2021**, *14*, 100503. [CrossRef] [PubMed]
35. Remington, B.C.; Westerhout, J.; Dubois, A.E.J.; Blom, W.M.; Kruizinga, A.G.; Taylor, S.L.; Houben, G.F.; Baumert, J.L. Suitability of low-dose, open food challenge data to supplement double-blind, placebo-controlled data in generation of food allergen threshold dose distributions. *Clin. Exp. Allergy* **2021**, *51*, 151–154. [CrossRef] [PubMed]
36. Holzhauser, T.; Johnson, P.; Hindley, J.P.; O'Connor, G.; Chan, C.-H.; Costa, J.; Fæste, C.K.; Hirst, B.J.; Lambertini, F.; Miani, M.; et al. Are current analytical methods suitable to verify VITAL® 2.0/3.0 allergen reference doses for EU allergens in foods? *Food Chem. Toxicol.* **2020**, *145*, 111709. [CrossRef]

Article

The Risk of Undeclared Allergens on Food Labels for Pediatric Patients in the European Union

Montserrat Martínez-Pineda * and Cristina Yagüe-Ruiz

Faculty of Health and Sports Science, University of Zaragoza, 22002 Huesca, Spain; cyague@unizar.es
* Correspondence: mmpineda@unizar.es; Tel.: +34-974-292759

Abstract: The dietary avoidance of allergens has been widely recognized as the key intervention in the management of food allergies, but the presence of undeclared allergens makes compliance difficult. The aim of this study was to analyze the presence of undeclared allergens in food labeling through RASFF notifications in the European Union, focusing on those allergens that frequently affect the pediatric population and the implicated products, so as to provide useful information for its risk evaluation and the development of educational materials for patients. The results showed milk (20.5%), gluten (14.8%), and nuts (10.9%) to be the pediatric allergens with higher presences. In 80% of the notifications concerning milk and milk derivatives, the specific compound present (lactose or lactoprotein) was not identified. They were mainly present in cereal and bakery products, prepared dishes and snacks, and cacao and confectionery products, all of which are frequently consumed by the pediatric population. The large quantity (7.6%) of undeclared allergens in "free-from-allergen" products was also remarkable, especially in regard to the supposedly not-present allergens. Undeclared allergens in food products pose an evident risk for allergic patients and knowledge of them should take a relevant role in a patient's nutritional education. It is also necessary to raise awareness among manufacturers and safety authorities.

Keywords: undeclared allergens; pediatric; food allergies; risk; RASFF

Citation: Martínez-Pineda, M.; Yagüe-Ruiz, C. The Risk of Undeclared Allergens on Food Labels for Pediatric Patients in the European Union. *Nutrients* **2022**, *14*, 1571. https://doi.org/10.3390/nu14081571

Academic Editor: Carla Mastrorilli

Received: 16 March 2022
Accepted: 8 April 2022
Published: 10 April 2022

Publisher's Note: MDPI stays neutral with regard to jurisdictional claims in published maps and institutional affiliations.

1. Introduction

The prevalence of allergic diseases worldwide is rising dramatically in both developed and developing countries. These diseases include asthma; rhinitis; anaphylaxis; drug, food, and insect allergies; eczema; urticaria (hives); and angioedema. This increase is especially problematic in children, who are bearing the greatest burden of the rising trend that has occurred over recent years [1]. Among allergic diseases, it is generally accepted that food allergies (FAs) affect approximately 2.5% of the general population, but the spread of prevalence data is wide, ranging from 1% to 19%, depending on patient age, diagnosis criteria, geographic area, etc. [2] In Europe, the estimated prevalence of FAs ranged from 1.0% to 5.6% in school-age children, while food sensitization (FS) ranged from 11.0% to 28.7%. Both primary and cross-reactive FS and FA occurred frequently at this age, according to data provided by Lyons et al. (2018) [3]

In children, the foods that frequently trigger allergic reactions include eggs, cow's milk, peanuts, tree nuts, soy, and wheat. Although some allergies typically resolve during childhood, allergies to peanuts and tree nuts, as well as those to fish and shellfish, remain into adulthood [4]. In children and adolescents <18 years of age, a systematic review that included 42 studies published in Europe between 2000 and 2012 reported a higher prevalence of food-challenge-defined allergies to cow's milk, 0.6% (0.5–0.8), followed by tree nuts, 0.5% (0.08–0.8), soy, 0.3% (0.1–0.4), eggs, 0.2% (0.2–0.3), peanuts, 0.2% (0.2–0.3), wheat, 0.1% (0.01–0.2), fish, 0.1% (0.02–0.2), and shellfish, 0.1% (0.06–0.3) although these percentages tended to be higher when allergy data were self-reported [5].

The clinical management of food allergies includes short-term interventions to manage acute reactions and long-term strategies to minimize the risk of further reactions. The strict dietary avoidance of allergens has been widely recognized as the key intervention in the management of FAs, resulting in the complete or almost complete resolution of symptoms [6]. In order to properly adhere to recommended elimination diets, patients and families should be instructed to pay careful attention to ingredient lists and food labels.

European legislation (EU Regulation No. 1169/2011) requires that information on the presence of allergens in foods is always provided to consumers, including on non-prepackaged foods. This regulation requires the indication of the presence of the 14 substances or products causing allergies or intolerances (hereinafter referred to as "allergens") shown in Table 1 when incorporated into food as ingredients [7]. This list was established on the basis of the scientific opinions adopted by the European Food Safety Authority (EFSA) [8].

Table 1. Mandatory declaration substances or products causing allergies or intolerances in the European Union according to EU Regulation 1169/2011 [7].

Substances or Products Causing Allergies or Intolerances	Exceptions
- Cereals containing gluten, namely wheat, rye, barley, oats, spelt, kamut, or their hybridized strains, and products thereof	- Wheat-based glucose syrups, including dextrose - Wheat-based maltodextrins - Glucose syrups based on barley - Cereals used for making alcoholic distillates, including ethyl alcohol, of agricultural origin
- Crustaceans and products thereof	
- Eggs and products thereof	
- Fish and products thereof	(a) Fish gelatin used as a carrier for vitamin or carotenoid preparations (b) Fish gelatin or isinglass used as a fining agent in beer and wine
- Peanuts and products thereof	
- Soybeans and products thereof	(a) Fully refined soybean oil and fat (b) Natural, mixed tocopherols (E306); natural D-alpha tocopherol; natural D-alpha tocopherol acetate; and natural D-alpha tocopherol succinate from soybean sources (c) Vegetable-oil-derived phytosterols and phytosterol esters from soybean sources (d) Plant stanol ester produced from vegetable-oil sterols from soybean sources
- Milk and products thereof (including lactose)	- Whey used for making alcoholic distillates, including ethyl alcohol, of agricultural origin - Lactitol
- Nuts, namely almonds (*Amygdalus communis* L.), hazelnuts (*Corylus avellana*), walnuts (*Juglans regia*), cashews (*Anacardium occidentale*), pecans (*Carya illinoinensis* (Wangenh.) K. Koch), Brazil nuts (*Bertholletia excelsa*), pistachio nuts (*Pistacia vera*), macadamia or Queensland nuts (*Macadamia ternifolia*), and products thereof, except for nuts used in making alcoholic distillates, including ethyl alcohol, of agricultural origin	
- Celery and products thereof	
- Mustard and products thereof	
- Sesame seeds and products thereof	
- Sulfur dioxide and sulfites at concentrations of more than 10 mg/kg or 10 mg/L in terms of the total SO_2, which are to be calculated for products proposed as ready for consumption or as reconstituted according to the instructions of the manufacturers	
- Lupin and products thereof	
- Mollusks and products thereof	

With regard to prepackaged foods, allergen information must appear in the list of ingredients, with clear references to the names of the substances or products given in Table 1. In addition, it should be highlighted by a typographical composition that clearly differentiates it from the rest of the list of ingredients (e.g., by typeface, style, or background color). In the absence of a list of ingredients, the word "contains" must be included, followed by the substance or product as listed in Table 1. An indication shall not be required in cases where the name of the food clearly refers to the substance or product causing allergies or intolerances.

With respect to non-prepackaged foods, the member states are allowed to adopt national measures concerning the means through which information on allergens on these foods is to be made available [7]. For example, in Spain, this is regulated by Royal Decree 126/2015 of 27 February 2015, which approves the general rule of food information on foodstuffs presented as unpackaged for sale to the final consumers and mass caterers, those packaged at the point of sale at the request of the purchaser, and those packaged by retail trade operators [9].

The food industry produces foods free of certain ingredients for consumers with food allergies or intolerances. In the European Union, there is legislation that regulates the requirements for the provision of information to consumers in the absence or reduced presence of gluten in food (EU Regulation No. 828/2014) [10]. This information should help gluten-intolerant people to identify and choose a varied diet when eating inside or outside their homes. There is also legislation regulating statements relating to the presence or absence of lactose in infant formula and follow-on formula (EU Regulation No. 2016/127), which can provide useful information to parents and caregivers [11]. However, there are still no harmonized rules at the EU level on labeling and composition indicating the absence or reduced presence of lactose in other foods. Given the importance of these claims for lactose-intolerant people, some member states have adopted non-binding guidelines. For example, the claims "lactose-free" and "low lactose" are used on foodstuffs for ordinary consumption marketed in Spain when the foodstuff contains less than 0.01% and 1% lactose, respectively [12].

The mandatory indication of allergenic compounds on labels is a very useful tool for patients to avoid consuming foods that contain allergens in their formulations. However, despite being mandatory, food mislabeling is on the rise and does not always adequately contain information about the allergens present [13], which implies a risk for allergic patients. Since the presence of undeclared allergens in food labeling has been considered to be a public health risk for a certain population, they have been included in the RASFF system [14]. The RASFF system database was created by the European Commission to keep the latest information on food recalls and public health warnings in all European Union (EU) countries, as well as Norway, Liechtenstein, Iceland, and Switzerland.

As a direct risk for allergic patients, the aim of the present study was to analyze the presence of undeclared allergens in food labeling through the European food-allergen-related notifications published on the RASFF portal from 2018 to 2021. Likewise, the analysis focused particularly on the undeclared presence of allergens that more frequently cause allergic reactions in the pediatric population, as well as on the food products that contain them, as useful information for patients' potential risk evaluation of commercial food products and as a relevant tool for the development of educational materials for families.

2. Materials and Methods

The data were obtained directly from the EU RASFF open data portal in .xlsx format [15]. The following items were available for each notification: date of notification; notifying country; origin country; type of product (food, food contact material, or feed); product category; product involved; hazard category; substance/finding; full hazard; subject; notification type; type of control; risk decision; distribution status; action taken after notification; and result (if available). From this online open database, notifications for the period between 1 January 2018 and 31 December 2021 were extracted under the hazard category filter "allergens" and type of product "food".

The following data were studied: date of notification; notification type; notifying country; origin country; type of control; main allergen hazard; product category; action taken after notification; and food product involved.

2.1. Main Allergen Hazard

This item refers to the allergen(s) involved in the notification. RASFF system data provide this information, grouping allergens according to the mandatory declaration allergens in prepackaged and non-prepackaged food products listed in Table 1. In some cases, the database provides specific information about the allergen, e.g., "wheat" or "oats" instead of "cereal containing gluten"; "lactose" or "lactoprotein" instead of "milk"; or a particular tree nut instead of "nuts". In the results section, these particular allergens were included and treated according to the mandatory declaration allergen groups to which they belong. Regarding crustaceans and mollusk allergens, when it was convenient for clarifying explanations, they were grouped and listed as "seafood allergen".

Furthermore, for all EU mandatory declaration allergens, the study paid particular attention to the analysis of those allergens of the greatest interest for the allergic pediatric population due to their higher prevalence.

2.2. Food Category

In addition to the main allergen hazard, the database gave a detailed description of the product involved in the notification (e.g., wafer rolls with cream). The product was also classified within one of the 27 food categories established by the RASFF system: alcoholic beverages; bivalve mollusks and products thereof; cephalopods and products thereof; cereals and bakery products; cocoa and cocoa preparations, coffee, and tea; confectionery; crustaceans and products thereof; dietetic foods, food supplements, and fortified foods; eggs and egg products; fats and oils; fish and fish products; food additives and flavorings; fruits and vegetables; gastropods; herbs and spices; honey and royal jelly; ices and desserts; meat and meat products (other than poultry); milk and milk products; natural mineral water; non-alcoholic beverages; nuts, nut products, and seeds; other food products/mixed; poultry meat and poultry-meat products; prepared dishes and snacks; soups, broths, sauces, and condiments; and wine.

2.3. Statistical Analysis

A descriptive statistical analysis of the data (proportions) was carried out for each of the items studied: notification type, action taken after notification, main allergen hazard, product category, and food product involved. Descriptive statistical analyses were performed with Microsoft Excel 2016® (Microsoft Corporation, Redmond, WA, USA).

A one-way ANOVA, followed by a Tukey post-hoc test, was applied to determine the statistical differences in the total number of undeclared allergen notifications among pediatric-relevant allergens (milk, gluten, nuts, soybean, egg, peanut, fish, crustaceans, and mollusks) realized in the period between 1 January 2018 and 31 December 2021. Values of $p < 0.05$ were considered significant. A two-way analysis of variance (ANOVA), followed by a Tukey post-hoc test, comparison test was performed to determine statistical differences in the total number of undeclared allergen notifications between years and allergens, as well as differences in the main pediatric allergens per food category. The same threshold for statistical significance ($p < 0.05$) was considered. These data were analyzed using GraphPad Prism version 6.01 (GraphPad Software, Inc., San Diego, CA, USA).

3. Results

Between 1 January 2018 and 31 December 2021, a total of 844 food-allergen-related notifications were made by the RASFF system, the year 2019 being the one with the highest number of notifications, $n = 241$, while the years 2018, 2020, and 2021 had $n = 207$, $n = 197$, and $n = 199$, respectively. It should be remarked that, in the period studied, 79.7% of the notifications were classified as an "alert", which implies that the food presents a serious risk on the market, requiring a rapid action generally aimed at withdrawing the product from the market. However, "information for attention and for follow-up" notifications represented only 17.4%. These types of notifications concern information that does not require a rapid action because the food product containing the undeclared allergen is still not on the market at the time of the report, or the risk is mostly considered low.

It is also relevant that, of the 844 allergen notifications reviewed, 16.6% corresponded to foods that contained two or more undeclared allergens, and could therefore potentially affect people with different allergies. Table 2 shows the proportion of each of them. It should be noted that several of these notifications revealed the presence of allergens that frequently affect children in products that easily could be consumed by them, for example, a 2018 notification that alerted about milk, soy, and wheat (in addition to mustard and celery) in organic beetroot soup.

Table 2. Percentage of RASFF notifications that informed about food products that simultaneously contained two or more undeclared allergens between 1 January 2018 and 31 December 2021.

Number of Allergens Present	% of Notifications
Two undeclared allergens	62.0
Three undeclared allergens	24.1
Four undeclared allergens	4.4
Five undeclared allergens	8.0
Six undeclared allergens	1.5

It can be observed that, from all those notifications, the main countries that emitted notifications about allergen hazards were Belgium, the Netherlands, and the United Kingdom. However, it should be remarked that 2021 data did not report notifications from the United Kingdom since it was no longer part of the EU as of January 2021. Regarding the countries of origin of the products concerned, it was observed that the main provenance was EU countries (77.6%), while 22.4% of notifications implied a non-EU/EEA country as the origin of the food.

From all of these notifications, it can be seen that the detection of the undeclared allergen in the food product came from a company's own check (51.1%), followed by official controls on the market (32.8%), and consumer complaints (9.6%). Notifications due to food poisoning represented 2.1% of the total notifications. These percentages remained very similar over the years studied.

In terms of responses to the notifications, the most frequently taken actions were the foods being recalled from the consumers (38.2%) and withdrawn from the market (19.9%) (Figure 1). All of these results are very relevant since they imply that the food products that contained the undeclared allergens were already on the market and could have already been consumed by an allergic person.

3.1. Main Allergen Hazard

According to the data, during the period studied, the main allergens about which notified were emitted were milk and products thereof, followed by cereals containing gluten and products thereof, sulfur dioxide and sulfites, and nuts (Figure 2), two of them (milk and nuts) being allergies with a higher prevalence in the pediatric population.

Apart from milk and its derivatives and nuts, other potential risk allergens in the pediatric population due to their high allergy prevalence, such as gluten, soy, eggs, or peanuts, were also involved in 36.4% of the notifications. On the other hand, it was found that the allergens that were the least frequently mentioned in the notifications were fish and seafood (crustaceans and mollusks), representing 1.5% and 1.9% of the notifications, respectively.

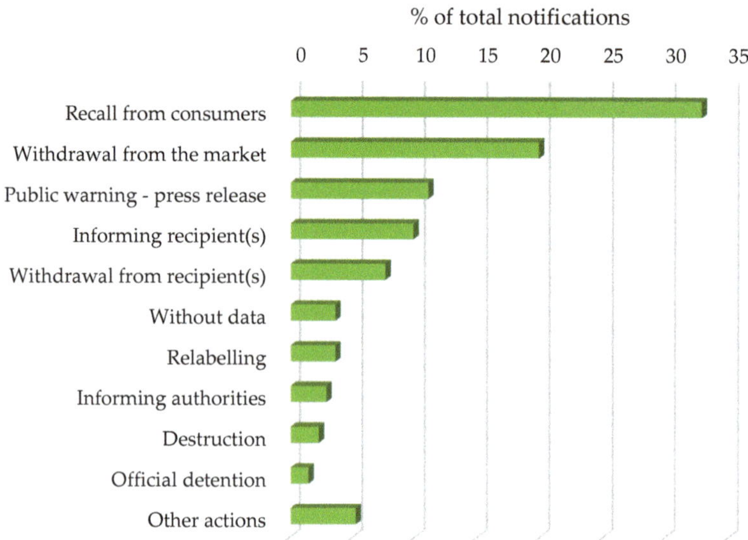

Figure 1. Percentage of actions taken as responses to notifications received between 1 January 2018 and 31 December 2021.

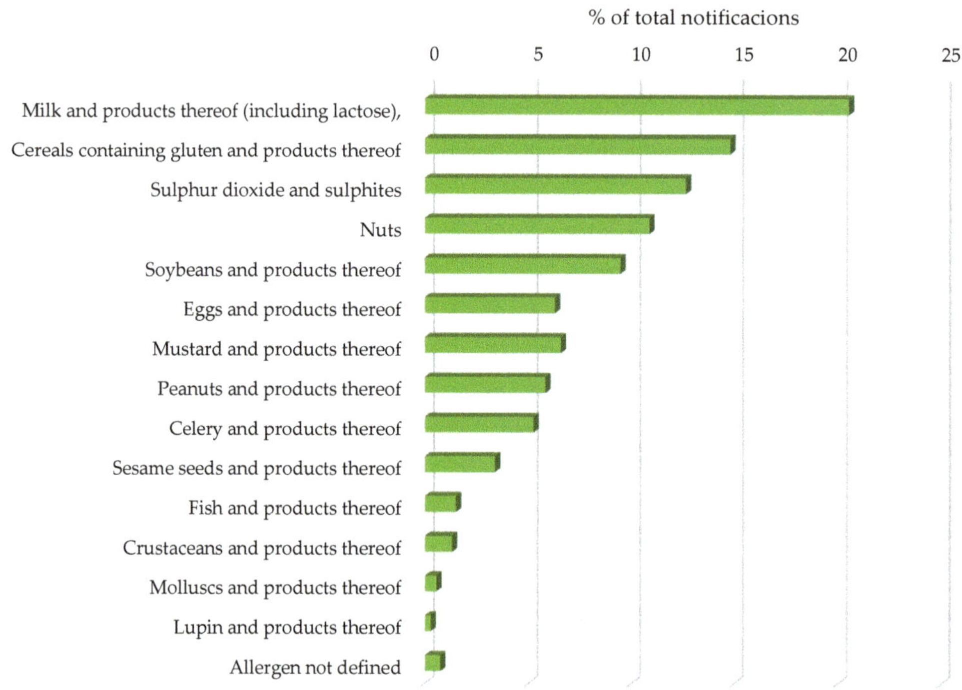

Figure 2. Percentage of total notifications per mandatory declaration allergen in the EU between 1 January 2018 and 31 December 2021.

The evolution of notifications-per-year of the main allergens that affect children and adolescents is shown in Figure 3. Globally, during the period studied, there were observed significant statistical differences between these allergen notifications (Figure 4); however, it should be remarked that, during the period of study, "milk and products thereof" remained the allergen that triggered the most notifications. Regarding differences in each allergen per year, the number of "undeclared milk notifications was significantly higher in 2019 and 2020, $p < 0.01$ and $p < 0.05$, respectively, compared to those of the year 2021, and regarding gluten notifications, the number was significantly higher ($p < 0.05$) in 2019 than in 2021. For the rest of the allergens, no significant differences were found over the four years.

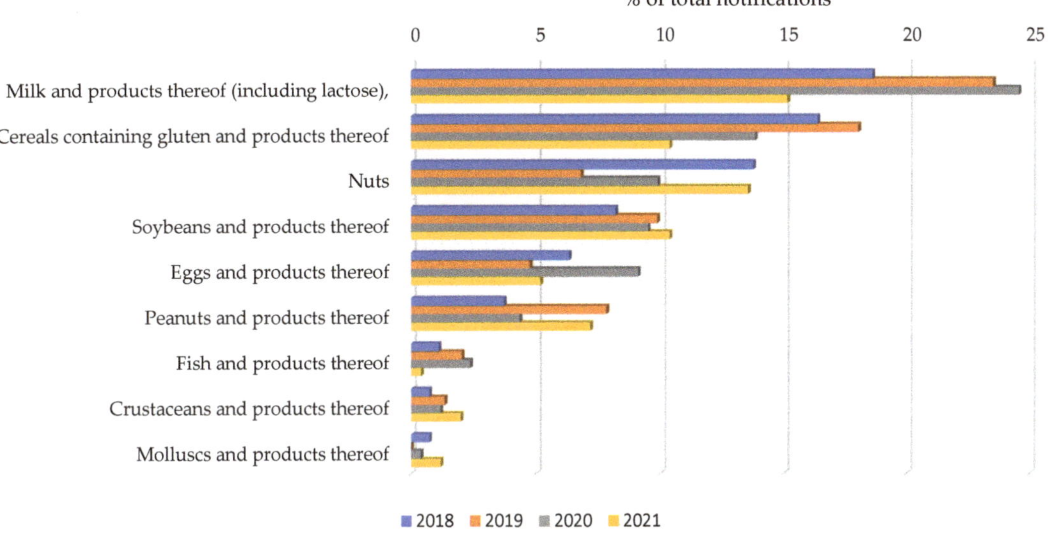

Figure 3. Percentage of total notifications of the main allergens affecting the pediatric population per year.

Figure 5 shows specific allergen notifications related to milk and nuts. In the case of milk and products thereof, only 20% of the related notifications specified the concrete compound present (lactose or lactoprotein), while the rest of the notifications only indicated the presence of milk. Conversely, 88% of the nut-related notifications specified the nuts involved; most of the notifications referred to the undeclared presence of almonds and hazelnuts. This information is relevant since a person could be lactose-intolerant but not necessarily allergic to a lactoprotein. If the response to a notification is a public warning (e.g., a press release), the use of the general term "milk" would not provide enough information to consumers.

In regard to almost all of the allergens that frequently affect the pediatric population, it was detected that the RASFF notifications were mainly due to labeling errors rather than the presence of the allergen or its traces due to possible cross-contamination. Among the labeling errors detected, it was also observed that some products that included information in several languages did not inform about the presence of allergens in one language but did in another.

The RASFF system included, in some cases, but not systematically, the quantity of the undeclared allergen in the product. With the available data, regarding milk and products thereof, it could be observed that the amount of undeclared milk (or milk ingredients) ranged between 0.35 mg/kg to 150 g/kg. The presence of lactoprotein in related notifications varied from 0.5 to 2500 mg/kg, and for undeclared lactose, quantities between 0.36 mg/kg and 15.6 g/kg were detected. In the case of undeclared peanuts, amounts between 1.2 mg/kg and 86.3 g/kg were found.

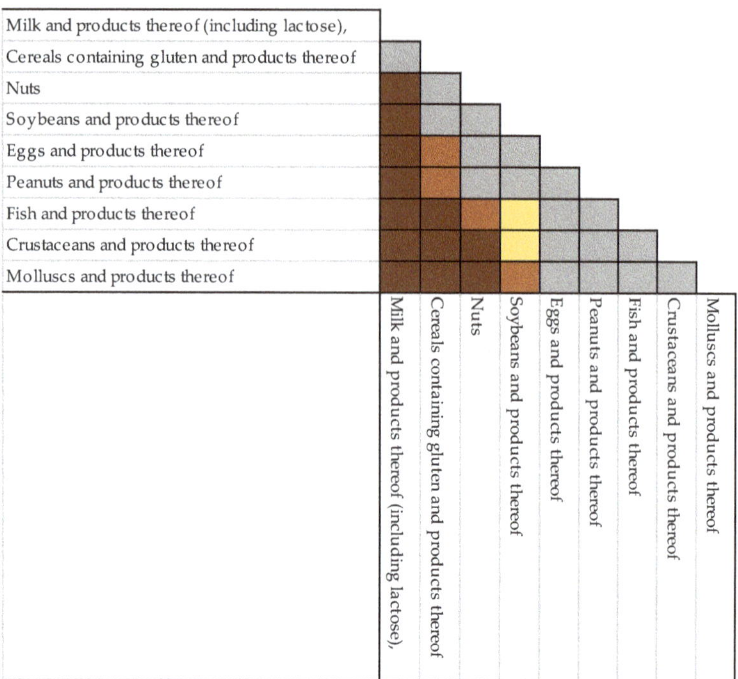

Figure 4. Statistical differences among the undeclared, main pediatric allergens in notifications emitted between 1 January 2018 and 31 December 2021. In grey- no significant differences; in yellow- significant differences with $p < 0.01$; in orange - significant differences with $p < 0.005$; and in brown-significant differences with $p < 0.001$.

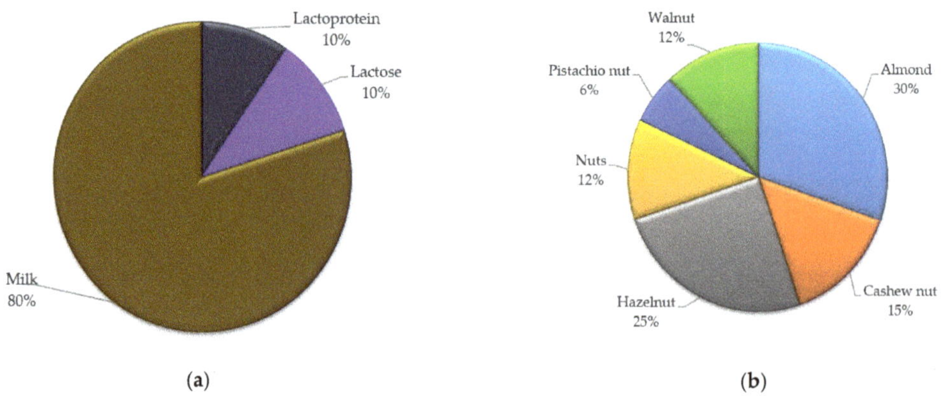

(**a**)　　　　　　　　　　　　　　　　(**b**)

Figure 5. Percentage of notifications related to undeclared milk and products thereof as well as nuts that specified the concrete compounds present in food products: (**a**) percentage of notifications that detailed milk-related allergens; (**b**) percentage of notifications that detailed nut-related allergens.

It should also be remarked that, for the undeclared gluten notifications, 23.2% of them were for gluten-free products due specifically to their gluten content being too high. According to EU Commission Implementing Regulation No. 828/2014 of 30 July 2014 on the requirements for the provision of information to consumers on the absence or reduced presence of gluten in food, the statement "gluten-free" may only be made where the food as sold to the final consumer contains no more than 20 mg/kg of gluten [10]. In all of these notifications, the gluten content exceeded that limit, being in some cases close to 1000 mg/kg. This quantity was as high as 1600 mg/kg in other products that were not labeled as "gluten-free," but which contained undeclared gluten.

3.2. Food Categories and Related Allergens

The notification percentage of each food category per year is shown in Figure 6. The results show that the main categories subject to notifications related to undeclared allergens were those of cereals and bakery products (16.3%), prepared dishes and snacks (13.1%), and confectionery (9.0%), followed by others that were classified as mixed products (e.g., frozen veggie burgers) (8.5%) and soups, broths, and sauces/condiments (7.7%). In general, those categories that include food products with a higher degree of processing and number of ingredients proved to be the ones that collected the highest number of notifications. The specific foods included in each category, according to what is stated in the RASFF database, are presented in the Supplementary Material (Table S1).

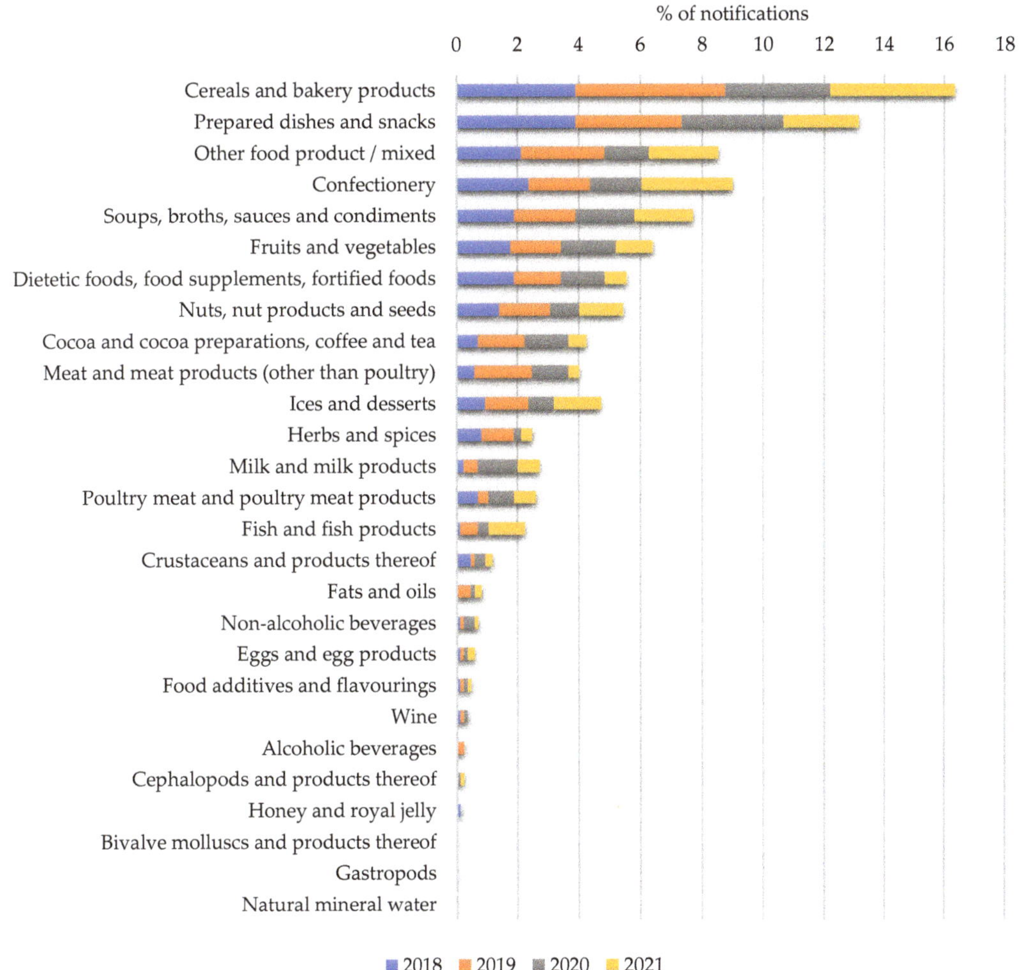

Figure 6. Percentage of notifications according to food categories per year.

Fruits and vegetables ranked sixth among the reported food categories. It should be clarified that the products involved were usually fruits and vegetables with different degrees of handling (sliced, grated, powder, pulp, purée, spread, and dried) or processing (dried, pickled, canned, and preserved), or vegetable mixtures.

Figure 7 shows the relationship between the main allergens affecting the pediatric population and their presence within each food category. The results revealed that soups, broths, sauces, and condiments, in addition to dietetic foods, food supplements, and fortified foods were the categories which included all common pediatric allergens (milk, nuts, gluten, egg, soy, peanut, fish, and seafood allergens). These categories were followed by prepared dishes and snacks as well as other mixed products which contained all allergens except peanut and seafood allergens, respectively. The types of food products included in the other/mixed products category are shown in Table A1 (Appendix A).

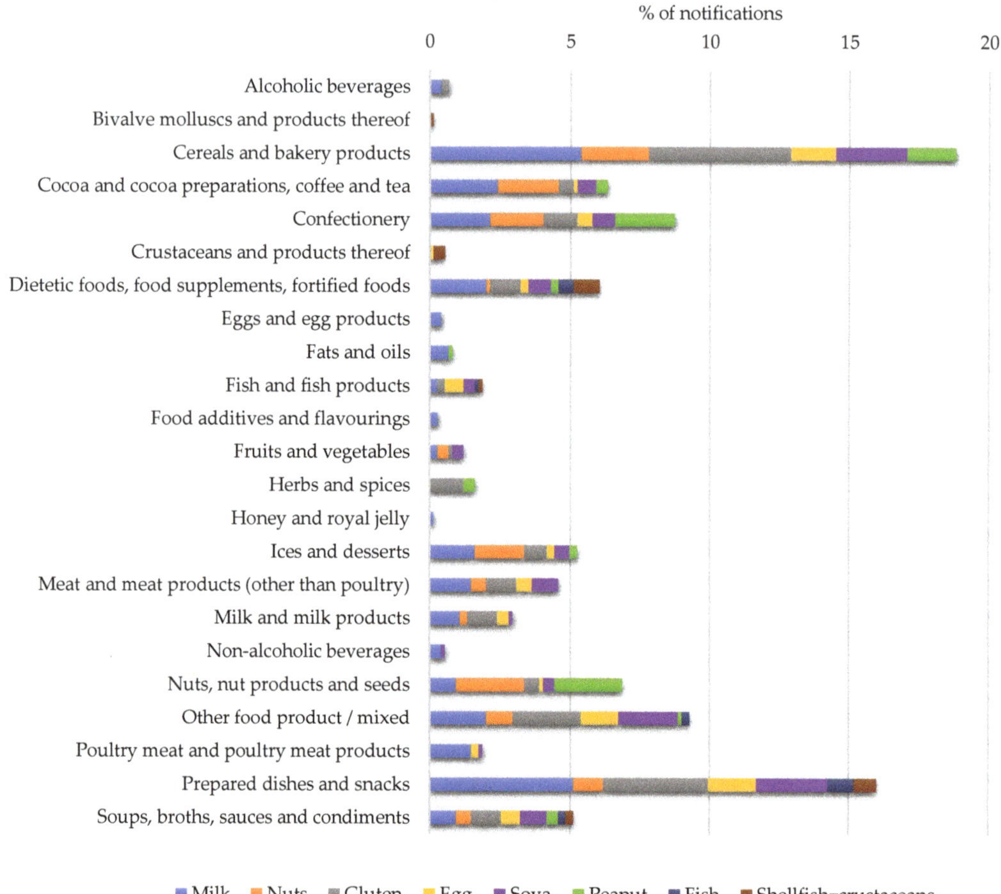

Figure 7. Relationship between the main allergens affecting the pediatric population and their presence within each food category.

The presence of undeclared milk was significantly higher ($p < 0.005$) in cereal and bakery products as well as in prepared dishes and snacks. Similar behavior was detected for gluten, with its undeclared presence also being significantly higher ($p < 0.001$ and $p < 0.05$, respectively) in these two food categories compared with the remainder oof the categories.

Regarding these two allergens, no significant differences in the number of notifications were observed among any of the other categories.

For the rest of the allergens, no significant statistical differences between the categories were observed. However, undeclared nuts were found predominantly in the cereal and bakery products category, nuts, nut products, and seeds, as well as cocoa and cocoa preparations. It should be noted that products included in these categories, especially the bakery and cocoa ones, are frequently ingested by infants.

For their parts, soy and eggs were mainly present in cereal and bakery products, as well as in prepared dishes and snacks. These results were expected since these foods are widely used as ingredients in both categories due to their technological functionalities, for example, as emulsifiers. Moreover, as both ingredients are widely used, the risk of cross-contamination within the industry is increased.

The presence of undeclared peanuts was higher in nuts, nut products, and seeds, as well as in the confectionary category. This can be easily explained due to cross-contamination because, although peanuts are legumes, their use and consumption are frequently linked with nut products (e.g., roasted, mixed nuts). Fish- and seafood-allergen-related notifications were mainly caused by dietetic foods and food supplements, but also by prepared dishes and snacks.

It should be noted that the cereal and bakery products category, in addition to being the one that caused the most notifications, warned of the undeclared presence of all of the most frequent allergens in children, except fish and seafood allergens. Most of the notifications in this food category were due to the presence of milk and products thereof, as well as cereals containing gluten.

It is especially noteworthy that 5.9% ($n = 50$) of the notifications were caused by gluten-free products. From them, 78% were due to the presence of gluten, while the remaining 22% were due to other allergens, such as lactose, milk, soy, or nuts, such as hazelnuts. The gluten-free products mainly involved were bars, biscuits, breads, cakes, chips, cookies, crisps, flours, granolas, muffins, pastas, noodles, nuggets, pizza toppings, ice creams, and desserts, among others. It should also be noted that many of them were also organic products. Additionally, other specially designed products for allergic or intolerant patients presented undeclared allergens, as can be observed in Table 3. These results are relevant since, in the case of allergic patients, their ingestion could directly affect their health status.

Table 3. Percentage of notifications which included specially designed products (gluten-free, lactose-free, dairy-free, etc.) and undeclared allergens involved in each case between 1 January 2018 and 31 December 2021.

Label Declaration	Total Notifications (%)	Undeclared Allergens	Examples of Notification-Related Products
"Gluten-free"	5.9	Gluten, lactose, milk, soy, and nuts	Organic, gluten-free, chocolate-coated, crispy cereals; gluten-free bread; and gluten-free pasta
"Dairy-free"	0.6	Milk, nuts, and soy	Chilled, dairy-free, coconut milk yogurt
"Lactose-free"	0.4	Milk, lactoprotein, lactose, and gluten	Lactose-free biscuits with buckwheat and chocolate
"Lactose-free and gluten-free"	0.4	Milk, lactoprotein, nuts, and soy	Gluten- and lactose-free chocolate spread
"Milk- and gluten-free"	0.2	Gluten, milk	Milk- and gluten-free vegan choco dessert with coconut cream
"Lactose-free" and "milk-protein-free"	0.1	Milk	Lactose- and milk-protein-free cake cream

In this context, products especially designed for covering alternative dietary patterns, such as those of vegetarians and vegans, which included statements such as "veggie", "vegetarian", or "vegan", represented 4.4% ($n = 37$) of the total notifications. Of them, 64.9% referred to undeclared allergens of animal origin (56.8% milk and 16.2% egg), while 35.1% were non-animal-related allergens (mustard, nuts, soy, celery, gluten, and peanuts).

Food products specifically intended for babies were also the subject of some notifications. This was the case for infant starter milk with the presence of fish allergens; baby cereal porridge with undeclared milk, lactose, and soya; and undeclared gluten in organic, gluten-free baby food.

4. Discussion

As there is no curative treatment for food allergies, and allergen avoidance is the mainstay of management, the presence of undeclared allergens in food implies a significantly dangerous risk for allergic patients. The avoidance of food allergens is onerous for patients and families and often fails, with 10% of patients on average experiencing at least one allergic reaction per year [16]. In their study, Fleischer et al. established that 87.4% of allergic reactions to foods in preschool-aged children were mainly caused by accidental exposure. Among causes of accidental reactions, unintentional ingestion (e.g., purely accidental due to forgetfulness, reduced supervision, not checking a product, etc.), label-reading errors, cross-contamination, errors in preparation, and manufacturer's labeling errors were found. The severity grade of these allergic reactions varied considerably, both among individuals and the allergens involved. Of all of the allergic reactions registered by the authors, 70.1% involved mild symptoms (skin, and/or oral symptoms, and/or upper-respiratory symptoms, but not all three organ systems); 18.4% involved moderate symptoms (skin, oral, upper-respiratory, or gastrointestinal symptoms); and 11.4% involved severe reactions (lower-respiratory symptoms; cardiovascular symptoms; or a combination of skin, oral, upper-respiratory, and gastrointestinal symptoms). These results reveal the impact of individual sensitivity to a particular allergen [17].

Although food-related anaphylaxis is relatively common, and all allergens are likely to cause severe reactions, fatalities remain rare, with a reported range of approximately 0.03 to 0.3 deaths per million people per year in the general population, and are very rare in infants and young children [18]. However, the vast majority of fatal allergic reactions were due to peanuts, tree nuts, seafood, and cow's milk [19]. Moreover, the consumption of non-prepackaged foods, served in catering establishments, self-service stores, bakeries, restaurants, etc., is frequently involved. This could probably be associated with a lack of direct allergen information on these kinds of products and the high risk of cross-contamination [20].

Despite the efforts made by the European Union to increase controls and consumers' information through labeling, including in community legislation regarding the mandatory declaration of allergens in food products, the results of the present study show that the risk due to the presence of undeclared allergens continues to be a problem for these patients. Previous studies have already highlighted that the presence of undeclared allergens was one of the main causes of food safety incidents/recalls [21]. An analysis of global recalls from previous years (from 2008 to 2018) also placed milk as the most frequently undeclared allergen, along with multiple allergens and gluten [22]. According to our results, this trend has persisted over time. However, contrary to what was observed in our results, nuts were not listed among the top allergens with the highest incidence in analyses of previous years and of non-EU countries [23,24]. Data from different countries and continents (RASFF vs. CDC, New Zealand, Australia, etc.), as well as the allergen-monitoring increase throughout the world in recent years, associated with the development of specific regulatory legislation, could explain the observed differences.

Previous studies also established cereal and bakery products as the most common food categories associated with the undeclared presence of milk, followed by confectionery [22]. However, this has changed over the four years of the study period since, from our results, it can be observed that milk allergens were more common in the recalls of prepared dishes than in those of confectionery. In any case, prepared dishes have already been identified by previous analyses as frequent causes of recalls [24]. The obtained results are worrying since both categories, derived from cereals and prepared dishes, are likely to be frequently consumed by children and adolescents, and they lead the list of products with the most prevalent allergens for this population.

FA patients often consume food products that self-report to be allergen-free (e.g., dairy-free) on their labels, trusting in their safety. However, the results exposed a concerning presence of undeclared allergens in a wide variety of those foods. This fact was especially striking in prepared meals and bakery products classified as "gluten-free". The evidence regarding the threshold limit of gluten concentration in food is also unclear. The consumption of about 200 mg gluten per day is clearly associated with the development of intestinal mucosal abnormalities after only 4 weeks in patients with celiac disease. However, it has been demonstrated that individual sensitivity to gluten varies among people with celiac disease, and the daily-intake limit should lie between 10 and 100 mg [25]. The length of exposure to gluten is also a determinant factor. For example, some authors have found that the ingestion of 10 or 50 mg gluten per day was associated with the worsening of the villous height/crypt depth ratio in most patients after 3 months [26]. In any case, under these premises, the high gluten amounts reported in some notifications (from 20 to more than 1600 mg/kg) highlight the obvious risk for some people with celiac disease and gluten intolerance, as the maximum limit ingestion must not exceed the 10 mg per day to avoid detrimental effects, such as significant histological abnormalities in some patients. According to European legislation, the statement "gluten-free" may only be made when the food as sold to the final consumer contains no more than 20 mg/kg of gluten, and the statement "very low gluten" may only be made where the food, consisting of or containing one or more ingredients made from wheat, rye, barley, oats, or their crossbred varieties, which have been specially processed to reduce the gluten content, contains no more than 100 mg/kg of gluten in the food as sold to the final consumer [10]. According to the painful limit of exposure, the serving portion of these products, and the quantity of gluten found, Table 4 shows examples of the potential allergen intake through the consumption of these products. It must be remarked that, in most of them, the estimated intake doses exceeded the limit of 10 mg in a serving portion.

Table 4. Examples of estimated gluten-intake dose via "gluten-free" products according to their serving portions and notifications of undeclared gluten amounts.

Product Claimed to be "Gluten-Free"	Quantity of Undeclared Gluten (mg/Kg—ppm)	Serving Portion (g)	Estimated Gluten Intake Dose (mg/Serving Portion)
Gluten-free corn pasta	176	150	26.4
Gluten-free cream preparation with potatoes and leeks	31.6	210	6.6
Gluten-free pasta	75.3	150	1.3
Gluten-free corn chips	265	30	8.0
Gluten-free peanut butter protein bar	853	40	34.1
Gluten-free hummus	>150	100	15.0
Gluten-free hemp protein	930	30	27.9
Gluten-free buckwheat and sweet potato noodles	>80	150	1.0
Gluten-free yellow lentil lasagna and spaghetti	88	200	17.6
Gluten-free vegetarian nuggets	38	100	3.8

The use of certain types of allergen-related labeling claims to attract consumers is a widespread practice among manufacturers, but a lack of control over potential allergens is often observed. In the case of patients allergic to milk or eggs, "vegan" products, such as soy or oat beverages, represent a suitable and safe alternative, such that they are frequently consumed by them. However, the results exposed that 64.9% of the notifications related to these types of products included undeclared allergens of animal origin (milk or eggs). The results are in accordance with those founded by Bedford et al. (2017). In their study, the authors observed that 50% of chocolates labeled "dairy-free" or "lactose-free", as well as 25% of those labeled "vegan", tested positive for milk, all with concentrations >1000 ppm [27].

Regarding milk as an allergen, current legislation only considers it mandatory to highlight the term "milk" to clearly differentiate it from the rest of a list of ingredients; however, it would be appropriate to review this legislation. When possible, labels should discern between the presence of lactose and milk proteins, instead of using only the term "milk" without more precise details; doing so would provide more practical information to patients, allowing them to make more convenient purchase choices according to their pathology, whether that be a milk protein allergy, lactose intolerance, or even if they are galactosemic. This discernment should also be systematically implemented in RASFF notifications to provide more accurate information for those patients. Although it varies a great deal according to the individual, as is the case with all food allergens, the threshold dose that induces symptoms in 5% of patients allergic to milk is less than 30 mg of milk proteins [28]. This low threshold dose implies an important risk for allergic consumers. It should also be remarked that, according to this symptomatic dose and the range of undeclared milk detected, a serving portion of 200 g of the product could imply, in some cases, a lactoprotein intake of 500 mg, almost 17 times higher than the dosage limit.

In contrast, regarding lactose, the EFSA Panel on Dietetic Products, Nutrition and Allergies concluded that symptoms of lactose intolerance have been described after the intake of less than 6 g of lactose in some subjects. However, the vast majority of subjects with lactose maldigestion could tolerate up to 12 g of lactose as a single dose with no (or only minor) symptoms. Additionally, it has been concluded that higher doses might be tolerated if they are distributed throughout the day [29]. It is relevant that even the highest amounts of undeclared lactose found (1.56 g/100 g) do not exceed the tolerance limit for intolerant patients. However, undeclared lactose also implies a serious risk for patients with galactosemia. For them, the threshold dose of lactose and galactose is lower, and strict lactose avoidance is required. It has been suggested to disallow all foods with a galactose content of >20 mg/100 g [30]. With all of this being the case, many of the undeclared lactose products about which notifications were emitted represented a severe danger for these patients.

In EU legislation, the claim "lactose-free" has only been defined for infant and follow-on formula ($\leq$10 mg/100 Kcal). Notwithstanding, some EU member states have set threshold levels at the national level for the use of the terms "lactose-free" or "low lactose" for foodstuffs other than those intended for infants. Unfortunately, a common level has not been adopted among these EU member states, and the lactose threshold level in "lactose-free" products varies between "absence of lactose and galactose" and 100 mg/100 g of the final product [29]. The lack of a common criterion adopted by all EU member states causes management and trade between countries difficult for manufacturers and can easily lead to notifications being emitted.

On the other hand, vegan patterns have globally risen during the last several years, including in FA pediatric patients [31,32]. Among the different motivations of plant-based dieters, the aversion to animal products due to moral and ethical reasons is highlighted [33]. In this context, the presence of undeclared allergens of animal origin, such as eggs, milk, fish, or seafood, in "vegan-claimed products" threatens the freedom of choice of this population and may pose a moral challenge to them.

Cross-reactivity is another problem of undeclared allergens in products. For example, a 75% cross-reactivity between soy as a primary food allergy and peanuts as a cross-reactive food has been observed [34]. If an allergic consumer were to ingest a product free of the primary trigger allergen, but it contains the cross-reactive undeclared allergen, it could imply a risk for their health.

The fact that 51.1% of the notifications of undeclared allergens had their origin in companies themselves highlights the efforts made by operators to control allergens. However, the proper management of allergens in the food industry implies a great challenge and a cost increase for them. Gupta et al. (2017) estimated this production cost increase as being between 10 and 30% [35]. To control this hazard, manufacturers routinely develop and implement independent allergen control plans to minimize the risk of product contact with food allergen contaminants and prevent recall events due to undeclared allergens. These plans typically specify practices for the safe handling and storage of raw materials, employee training, facility and equipment design, cleaning procedures, and production scheduling. Recalls due to food allergen cross-contact, cleaning procedures, equipment and premises design, and employee training were ranked by companies as the greatest allergen management expenses. In addition, companies may use precautionary allergen labeling (PAL), such as "may contain" on packaging, to label products for which there is a risk of cross-contact with food allergens during production. However, PAL usage remains voluntary and unregulated, and it currently presents consumers with considerable challenges due to its inconsistent use. The European Commission should also direct its efforts to address the inconsistent usage of PALs, promoting the harmonization of language used in PALs, and improving PAL status to quantified PAL statements, as previously suggested by other authors [36]. It would be helpful in communicating risks for both manufacturers and consumers, so that they can make informed choices when purchasing food products.

Food producers must continue increasing their efforts to improve food allergen management in order to reduce the presence of undeclared allergens in their products. For example, to reach this objective, they can follow and apply available guidance, such as *Guidance on Food Allergen Management for Food Manufacturers* [37]. Furthermore, it would also be highly recommended to implement international standards, such as voluntary food security and quality certifications, for example, International Food Standards version 6.1 [38] or BRGCS's Gluten-Free Certification Program (GFCP) [39]. Currently, scientific advances have allowed the development of technological alternatives, such as irradiation or high hydrostatic pressure, in which, in addition to preserving nutrients, freshness, and organoleptic characteristics, they alter the structure of the proteins causing allergies, reducing their allergenicity [40,41]. However, these kinds of technologies are still under development, and their costs are very high, so they cannot be used by many companies. In the current scene, food manufacturers should improve food allergen management in their practices by focusing on empowering employees through more knowledge about food allergens and allergies, as well as through the use of new digital tools such as big data, as previously proposed by Jia and Evans (2021) [42].

The management of food allergies and dietary avoidance presents several challenges for pediatric dietitians and other healthcare providers [43]. Health professionals who assess FA patients should include nutrition therapy to ensure the adequate intake of nutrients as well as nutritional education with comprehensive information about allergenic ingredients for their avoidance [31,44]. The American Academy of Pediatrics established as a critical issue the improving of the education and training of all stakeholders for recognizing and managing, as well as preventing, allergic reactions. This work includes updating, creating, and implementing various guidelines and educational programs [45]. In this line, educational programs should be based on three basic aspects:

1. Learning which foods, due to their composition, may contain the trigger allergen.

2. Understanding how to read labels to find out if the trigger allergen is present, including "contains traces of" or "may contain" statements. In this sense, it is also essential to note that unpackaged products, packaged directly in store or those sold in the form of self-service (e.g., bread and pastries offered in supermarkets) are very prone to cross-contamination with various allergens. However, despite it being mandatory to inform customers about the allergens present, the absence of a label on products and the high risk of cross-contamination cause their consumption by allergic patients to be unrecommended.

3. Learning resources containing information about undeclared allergens in food labeling, as well as those products more likely to contain them according to the trigger allergen. On this point, it would be useful to provide information about official resources and webpages that share food allergen alerts, such as RASFF.

It would be helpful to provide patients with summarized and visual information about those products that most frequently present undeclared allergens, such as that shown in Table 5.

Table 5. Summary of the main allergens that affect the pediatric population, and main risk food category.

Allergen	Notified Food Category	Notified Food Examples
Milk (lactoproteins and lactose)	Cereal and bakery products	Organic rice pancakes with dark chocolate, biscuits, and popcorn
	Prepared dishes and snacks	Sausage rolls, grilled chips, and packed snacks
Nuts	Cereal and bakery products	Pesto-and-walnut-plucked bread, chocolate pie, and energy cereal bars
	Nuts, nut products, and seeds	Pistachio cream; fruit–nut mix; and peanut, nut, and mulberry mix
	Cocoa and cocoa preparations	Milk chocolate with buckwheat; chocolate; and chocolate spread
Gluten	Cereal and bakery products	Rice-flour cake mix, sugar loaves, bulgur and pasta, and hot dog buns
	Prepared dishes and snacks	Rice salad, baby food, and tortilla chips
Egg	Cereal and bakery products	Donuts with cocoa coating, red velvet muffins, and bakery scones
	Prepared dishes and snacks	Hummus and aioli, pizza, and prepared sandwiches
Soya	Cereal and bakery products	Rice flour, cheese and onion bread, and butter croissants
	Prepared dishes and snacks	Beef-flavored instant rice noodles, nacho-cheese snacks, and potato chips
Peanut	Nuts, nut products, and seeds	Almond paste; nut mix; and roasted, organic, almond kernels
	Confectionery	Baklava, wafer rolls with cream, and candies
Fish	Dietetic foods, food supplements, and fortified foods	Food supplements and organic infant milk starter packs
	Prepared dishes and snacks	Club-salad pasta, tuna and chicken, chilled tandoori chicken salad dishes, and salmon lasagna
Seafood (mollusks and crustaceans)	Dietetic foods, food supplements, and fortified foods	Food supplements
	Prepared dishes and snacks	Frozen cheese croquettes; frozen lamb and carrot dumplings; and chilled, spicy chicken salad

As it has been demonstrated that labeling is not a sufficient guarantee, and that there may be undeclared allergens, a useful reference tool for allergic patients and their families is the RASFF consumers' portal. Launched in June 2014, this website is a consumer-friendly internet tool that provides the latest information on food recall notices. It also includes public health warnings issued by food safety authorities and food companies. By using this free tool, EU consumers could identify food that has been flagged in the system, allowing them to make more safety choices. There, users obtain access to practical information on product recalls and public health warnings in any given EU country. Additionally, they could also consult the institutional websites of the member states. The RASFF consumers' portal and most websites provide food alert information on any food hazards, including undeclared allergens. Some of them are shown in Table 6.

Table 6. Examples of institutional websites of European Union countries that provide practical undeclared allergen information (product recalls and other public health warnings).

Country	Institution/ Public Organism	Website	
European Commission	RASFF consumers' portal	https://webgate.ec.europa.eu/rasff-window/screen/consumers (accessed on 10 March 2022)	[14]
Austria	Austrian Agency for Health and Food Safety (AGES)	https://www.ages.at/en/human/product-warnings-product-recalls (accessed on 10 March 2022)	[46]
Belgium	Federal Agency for the Safety of the Food Chain (AFSCA)	https://www.favv-afsca.be/consommateurs/avertissements/ (accessed on 10 March 2022)	[47]
Croatia	Government of the Republic of Croatia	https://gov.hr/en/hrana-food-related-warnings/1774 (accessed on 10 March 2022)	[48]

Table 6. *Cont.*

Country	Institution/ Public Organism	Website	
France	Ministry of Agriculture and Food	https://rappel.conso.gouv.fr/categorie/1#navigation (accessed on 10 March 2022)	[49]
Germany	Federal Office of Consumer Protection and Food Safety	https://www.lebensmittelwarnung.de/bvl-lmwde/liste/lebensmittel/deutschlandweit/10/0 (accessed on 10 March 2022)	[50]
Greece	Hellenic Food Authority (EFET)	https://www.efet.gr/index.php/en/ (accessed on 10 March 2022)	[51]
Ireland	Food Safety Authority of Ireland	https://www.fsai.ie/ (accessed on 10 March 2022)	[52]
Italy	Ministry of Health	https://www.salute.gov.it/portale/news/p3_2_1_3.jsp?lingua=italiano&menu=notizie&p=avvisi (accessed on 10 March 2022)	[53]
Luxembourg	Food safety—Grand Duchy of Luxembourg	https://securite-alimentaire.public.lu/fr.html (accessed on 10 March 2022)	[54]
Malta	Environmental Health Directorate	https://deputyprimeminister.gov.mt/en/environmental/Pages/Home-Page.aspx (accessed on 10 March 2022)	[55]
Netherlands	Netherlands Food and Consumers Product Safety Authority	https://www.nvwa.nl/onderwerpen/veiligheidswaarschuwingen (accessed on 10 March 2022)	[56]
Romania	National Sanitary Veterinary and Food Safety Authority (ANSVSA)	http://www.ansvsa.ro/informatii-pentru-public/produse-rechemateretrase/ (accessed on 10 March 2022)	[57]
Slovenia	Republic of Slovenia	https://www.gov.si/drzavni-organi/organi-v-sestavi/uprava-za-varno-hrano-veterinarstvo-in-varstvo-rastlin/ (accessed on 10 March 2022)	[58]
Spain	Spanish Agency for Food Safety and Nutrition (AECOSAN)	https://www.aesan.gob.es/AECOSAN/web/subhomes/seguridad_alimentaria/aecosan_seguridad_alimentaria.htm (accessed on 10 March 2022)	[59]

In this respect, free mobile applications, such as HRana from the Croatian Ministry of Agriculture, are very useful for consumers in general, and for allergic patients in particular. This application enables citizens to receive information within 24 h on warnings regarding food, animal feed, and objects, as well as materials that come into direct contact with food which are sold in the Croatian and/or EU markets. Information that the Ministry of Agriculture updates in real time is sent to citizens via an application notification that displays food-related warnings and additional information on the non-compliant product (product name and image, shelf life, type of health risk, action taken by competent institutions, distribution status, etc.), as well as information on the food business operator that markets the product [60].

5. Conclusions

The results of this study reinforce the evidence of the risk associated with the presence of undeclared allergens on food labels for allergic patients. Of particular concern is the undeclared presence of those allergens with the highest prevalence in the pediatric population, in foods frequently consumed by this population, such as cereal and bakery products, prepared dishes and snacks, and cocoa and confectionery. Manufacturers and safety authorities should increase their efforts to face this risk, and educational programs could enforce patients' knowledge about potential, undeclared-allergen food products with official resources and webpages, such as the RASFF consumers' portal, which share updated food allergen alerts.

Supplementary Materials: The following are available online at https://www.mdpi.com/article/10.3390/nu14081571/s1. Table S1: Food products included per food category in the RASFF notifications from 2018 to 2021.

Author Contributions: Conceptualization: C.Y.-R.; methodology: C.Y.-R.; formal analysis: M.M.-P. and C.Y.-R.; investigation: M.M.-P. and C.Y.-R.; resources: C.Y.-R.; data curation: M.M.-P.; writing—original draft: M.M.-P. and C.Y.-R.; writing—review and editing: M.M.-P. and C.Y.-R.; visualization: M.M.-P.; supervision: C.Y.-R.; funding acquisition: C.Y.-R.; project administration: C.Y.-R. All authors have read and agreed to the published version of the manuscript.

Funding: The APC was funded by the Government of Aragón (grant Grupo A06-20R).

Institutional Review Board Statement: Not applicable.

Informed Consent Statement: Not applicable.

Conflicts of Interest: The authors declare no conflict of interest.

Appendix A

Table A1. Food products involved in "other/mixed products category" in the RASFF notifications from 1 January 2018 and 31 December 2021.

Year 2018	Year 2019	Year 2020	Year 2021
Aubergines in oil	Carbonated strawberry-and-cream-flavored soft drink	Sweet and sour chicken	Canned soybeans mislabeled as spelt
Cake cream labelled as "lactose-free" and "milk-protein-free"	Carrot and dill hummus		Chilled, smoked-salmon salads
Canned, peeled tomatoes	Chilled Hawaiian chicken salad	Frozen, organic, mung-bean nuggets	Chilled strawberries with chocolate cream
Chips	Frozen garlic puree	Frozen spring-roll sheets	Chilled vegan paprika slices
Cooked white beans	Frozen potato waffles	Frozen veggie burgers	Croutons used in Caesar salad
Easter-egg dye	Frozen soya product	Melba toast	Frozen beef spring rolls
Gluten-free bread mix	Frozen spicy burgers	Organic coconut sugar	Frozen sandwiches
Gluten-free-protein linseed and protein powder	Gluten-free hemp protein	Roasted vegetables	Grilled bell-pepper tapenade
Grilled chicken dish and tuna salad	Gluten-free hummus	Spinach noodles	Olives
Mini mooncakes	Green olives in glass jars	Various frozen products	Organic, gluten-free, soy flour
Organic, veggie-style chicken chunks mislabeled and packed as veggie-style beef strips	Grilled artichokes in sunflower oil	Vegan pate	Organic hummus
Pickled pepperoni	Instant pumpkin cereal	Vegetarian, marinated fillet	Pink roses for cake decoration
Various foodstuffs	Mini vegetarian-burger nuggets		Spaghetti Bolognese
Various halawa flavors and jams	Olive paste		Spicy peppers filled with tuna and capers
Vegan, dairy-free, grated pizza topping	Organic powder preparation for fermented soy dessert		Spinach cream
Vegan, stracciatella-flavored, lupin-based yogurt	Pasta salad with smoked salmon		Spreads
Wafer sheets	Seitan and vegetable sausages		Various gluten-free, processed products
	Soy-meat products		Vegetarian burgers
	Spicy carrot spread with ginger		Vegetarian sausages
	Sugar-free licorice sweets		
	Vegetable snack		
	Vegetarian mince mistakenly packaged as roasted cubes		
	Vegetarian salami		

References

1. World Health Organization; Canonica, G.W.; Holgate, S.; Lockey, R.; Pawankar, R. *White Book on Allergy 2011–2012 Executive Summary*; World Allergy Organization: Milwaukee, WI, USA, 2013.
2. Loh, W.; Tang, M.L.K. The Epidemiology of Food Allergy in the Global Context. *Int. J. Environ. Res. Public Health* **2018**, *15*, 2043. [CrossRef] [PubMed]
3. Lyons, S.A.; Clausen, M.; Knulst, A.C.; Ballmer-Weber, B.K.; Fernandez-Rivas, M.; Barreales, L.; Bieli, C.; Dubakiene, R.; Fernandez-Perez, C.; Jedrzejczak-Czechowicz, M.; et al. Prevalence of Food Sensitization and Food Allergy in Children across Europe. *J. Allergy Clin. Immunol. Pract.* **2020**, *8*, 2736–2746.e9. [CrossRef] [PubMed]
4. Sicherer, S.H.; Sampson, H.A. Food Allergy: Epidemiology, Pathogenesis, Diagnosis, and Treatment. *J. Allergy Clin. Immunol.* **2013**, *133*, 291–307.e5. [CrossRef] [PubMed]
5. Nwaru, B.I.; Hickstein, L.; Panesar, S.S.; Roberts, G.; Muraro, A.; Sheikh, A. Prevalence of Common Food Allergies in Europe: A Systematic Review and Meta-Analysis. *Allergy* **2014**, *69*, 992–1007. [CrossRef] [PubMed]
6. Agache, I. *EAACI White Paper on Research, Innovation and Quality Care*; European Academy of Allergy and Clinical Immunology: Zurich, Switzerland, 2018.
7. The European Parliament and the Council of the European Union. Regulation (EU) No 1169/2011 of the European Parliament and of the Council of 25 October 2011 on the Provision of Food Information to Consumers, Amending Regulations (EC) No 1924/2006 and (EC) No 1925/2006 of the European Parliament and of the Council, and Repealing Commission Directive 87/250/EEC, Council Directive 90/496/EEC, Commission Directive 1999/10/EC, Directive 2000/13/EC of the European Parliament and of the Council, Commission Directives 2002/67/EC and 2008/5/EC and Commission Regulation (EC) No 608/2004 Text with EEA Relevance. *Off. J. Eur. Union* **2011**, *304*, 1–65.
8. EFSA NDA Panel (EFSA Panel on Dietetic Products, Nutrition and Allergies). Scientific Opinion on the Evaluation of Allergenic Foods and Food Ingredients for Labelling Purposes. *EFSA J.* **2014**, *12*, 3894.
9. Royal Decree 126/2015 of 27 February 2015, which Approves the General Rule on Food Information on Foodstuffs Presented Unpackaged for Sale to the Final Consumer and to Mass Caterers, those Packaged at the Point of Sale at the Request of the Purchaser, and those Packaged by Retail Trade Operators. *BOE* **2015**, *54*, 20059–20066.
10. Commission Implementing Regulation (EU) no 828/2014 of 30 July 2014 on the Requirements for the Provision of Information to Consumers on the Absence Or Reduced Presence of Gluten in Food (Text with EEA Relevance). *Off. J. Eur. Union* **2014**, *228*, 5–8.
11. Commission Delegated Regulation (EU) 2016/127 of 25 September 2015 Supplementing Regulation (EU) no 609/2013 of the European Parliament and of the Council as Regards the Specific Compositional and Information Requirements for Infant Formula and Follow-on Formula and as Regards Requirements on Information Relating to Infant and Young Child Feeding (Text with EEA Relevance). *Off. J. Eur. Union* **2016**, *25*, 1–29.
12. AESAN. Notice of 26 July 2019 Relating to the Conditions of use of the Terms 'Lactose-Free' and 'Low Lactose'. Available online: https://www.aesan.gob.es/AECOSAN/docs/documentos/seguridad_alimentaria/interpretaciones/nutricionales/sin_lactosa.pdf (accessed on 2 April 2022).
13. Soon, J.M.; Abdul Wahab, I.R. Global Food Recalls and Alerts Associated with Labelling Errors and its Contributory Factors. *Trends Food Sci. Technol.* **2021**, *118*, 791–798. [CrossRef]
14. RASFF Window. Version 2.1.0. European Commission. Available online: https://webgate.ec.europa.eu/rasff-window/screen/search (accessed on 2 January 2022).
15. The Official Portal for European Data. European Commission. Available online: https://data.europa.eu/en (accessed on 2 January 2022).
16. Peters, R.L.; Krawiec, M.; Koplin, J.J.; Santos, A.F. Update on Food Allergy. *Pediatr. Allergy Immunol.* **2021**, *32*, 647. [CrossRef] [PubMed]
17. Fleischer, D.M.; Perry, T.T.; Atkins, D.; Wood, R.A.; Burks, A.W.; Jones, S.M.; Henning, A.K.; Stablein, D.; Sampson, H.A.; Sicherer, S.H. Allergic Reactions to Foods in Preschool-Aged Children in a Prospective Observational Food Allergy Study. *Pediatrics* **2012**, *130*, e25–e32. [CrossRef]
18. Turner, P.J.; Jerschow, E.; Umasunthar, T.; Lin, R.; Campbell, D.E.; Boyle, R.J. Fatal Anaphylaxis: Mortality Rate and Risk Factors. *J. Allergy Clin. Immunol. Pract.* **2017**, *5*, 1169–1178. [CrossRef] [PubMed]
19. Baseggio Conrado, A.; Patel, N.; Turner, P.J. Global Patterns in Anaphylaxis due to Specific Foods: A systematic Review. *J. Allergy Clin. Immunol.* **2021**, *148*, 1515–1525.e3. [CrossRef] [PubMed]
20. Turner, P.J.; Gowland, M.H.; Sharma, V.; Ierodiakonou, D.; Harper, N.; Garcez, T.; Pumphrey, R.; Boyle, R.J. Increase in Anaphylaxis-Related Hospitalizations but no Increase in Fatalities: An Analysis of United Kingdom National Anaphylaxis Data, 1992–2012. *J. Allergy Clin. Immunol.* **2014**, *135*, 956–963.e1. [CrossRef]
21. Kleter, G.A.; Prandini, A.; Filippi, L.; Marvin, H.J.P. Identification of Potentially Emerging Food Safety Issues by Analysis of Reports Published by the European Community's Rapid Alert System for Food and Feed (RASFF) during a Four-Year Period. *Food Chem. Toxicol.* **2009**, *47*, 932–950. [CrossRef]
22. Soon, J.M.; Brazier, A.K.M.; Wallace, C.A. Determining Common Contributory Factors in Food Safety Incidents—A Review of Global Outbreaks and Recalls 2008–2018. *Trends Food Sci. Technol.* **2020**, *97*, 76–87. [CrossRef]
23. Gendel, S.M.; Jianmei, Z. Analysis of U.S. Food and Drug Administration Food Allergen Recalls After Implementation of the Food Allergen Labeling and Consumer Protection Act. *J. Food Prot.* **2013**, *76*, 1933–1938. [CrossRef]

24. Bucchini, L.; Guzzon, A.; Poms, R.; Senyuva, H. Analysis and Critical Comparison of Food Allergen Recalls from the European Union, USA, Canada, Hong Kong, Australia and New Zealand. *Food Addit. Contam. Part A Chem. Anal. Control Expo. Risk Assess.* **2016**, *33*, 760–771. [CrossRef]

25. Hischenhuber, C.; Crevel, R.; Jarry, B.; Maki, M.; Moneret-Vautrin, D.A.; Romano, A.; Troncone, R.; Ward, R. Review Article: Safe Amounts of Gluten for Patients with Wheat Allergy Or Coeliac Disease. *Aliment Pharmacol. Ther.* **2006**, *23*, 559–575. [CrossRef] [PubMed]

26. Akobeng, A.K.; Thomas, A.G. Systematic Review: Tolerable Amount of Gluten for People with Coeliac Disease. *Aliment Pharmacol. Ther.* **2008**, *27*, 1044–1052. [CrossRef] [PubMed]

27. Bedford, B.; Yu, Y.; Wang, X.; Garber, E.A.E.; Jackson, L.S. A Limited Survey of Dark Chocolate Bars obtained in the United States for Undeclared Milk and Peanut Allergens. *J. Food Prot.* **2017**, *80*, 692–702. [CrossRef] [PubMed]

28. Moneret-Vautrin, D.A.; Kanny, G. Update on Threshold Doses of Food Allergens: Implications for Patients and the Food Industry. *Curr. Opin. Allergy Clin. Immunol.* **2004**, *4*, 215–219. [CrossRef] [PubMed]

29. EFSA Panel on Dietetic Products, Nutrition and Allergies (NDA). Scientific Opinion on Lactose Thresholds in Lactose Intolerance and Galactosaemia. *EFSA J.* **2010**, *8*, 1777. [CrossRef]

30. Gropper, S.; Weese, J.; West, P.; Gross, K. Free Galactose Content of Fresh Fruits and Strained Fruit and Vegetable Baby Foods:More Foods to Consider for the Galactose-Restricted Diet. *J. Acad. Nutr. Diet.* **2000**, *100*, 573–575.

31. Protudjer, J.L.P.; Mikkelsen, A. Veganism and Paediatric Food Allergy: Two Increasingly Prevalent Dietary Issues that are Challenging when Co-Occurring. *BMC Pediatr.* **2020**, *20*, 341. [CrossRef]

32. Sutter, D.O.; Bender, N. Nutrient Status and Growth in Vegan Children. *Nutr. Res.* **2021**, *91*, 13–25. [CrossRef]

33. Rosenfeld, D.L.; Burrow, A.L. Vegetarian on Purpose: Understanding the Motivations of Plant-Based Dieters. *Appetite* **2017**, *116*, 456–463. [CrossRef]

34. Cox, A.L.; Eigenmann, P.A.; Sicherer, S.H. Clinical Relevance of Cross-Reactivity in Food Allergy. *J. Allergy Clin. Immunol. Pract.* **2021**, *9*, 82–99. [CrossRef]

35. Gupta, R.S.; Taylor, S.L.; Baumert, J.L.; Kao, L.M.; Schuster, E.; Smith, B.M. Economic Factors Impacting Food Allergen Management: Perspectives from the Food Industry. *J. Food Prot.* **2017**, *80*, 1719. [CrossRef]

36. Soon, J.M.; Manning, L. "May Contain" Allergen Statements: Facilitating Or Frustrating Consumers? *J. Consum. Policy* **2017**, *40*, 447–472. [CrossRef]

37. Food Drink Europe. *Guidance on Food Allergen Management for Food Manufacturers*; Food Drink Europe: Brussels, Belgium, 2013; 88p.

38. IFS Management GmbH. *IFS Food Version 6.1.*; IFS Management GmbH: Berlin, Germany, 2017; 158p.

39. BRCGS. *Gluten-Free Certification Program Global Standard (Issue 3)*; BRCGS: London, UK, 2019.

40. Zhang, Y.; Ren, Y.; Bi, Y.; Wang, Q.; Cheng, K.; Chen, F. Review: Seafood Allergy and Potential Application of High Hydrostatic Pressure to Reduce Seafood Allergenicity. *Int. J. Food Eng.* **2019**, *15*, 20180392. [CrossRef]

41. Pan, M.; Yang, J.; Liu, K.; Xie, X.; Hong, L.; Wang, S.; Wang, S. Irradiation Technology: An Effective and Promising Strategy for Eliminating Food Allergens. *Food Res. Int.* **2021**, *148*, 110578. [CrossRef] [PubMed]

42. Jia, L.; Evans, S. Improving Food Allergen Management in Food Manufacturing: An Incentive-Based Approach. *Food Control* **2021**, *129*, 108246. [CrossRef]

43. Groetch, M.E.; Christie, L.; Vargas, P.A.; Jones, S.M.; Sicherer, S.H. Food Allergy Educational Needs of Pediatric Dietitians: A Survey by the Consortium of Food Allergy Research. *J. Nutr. Educ. Behav.* **2010**, *42*, 259–264. [CrossRef]

44. Collins, S.C. Practice Paper of the Academy of Nutrition and Dietetics: Role of the Registered Dietitian Nutritionist in the Diagnosis and Management of Food Allergies. *J. Acad. Nutr. Diet.* **2016**, *116*, 1621–1631. [CrossRef]

45. Sicherer, S.H.; Allen, K.; Lack, G.; Taylor, S.L.; Donovan, S.M.; Oria, M. Critical Issues in Food Allergy: A National Academies Consensus Report. *Pediatrics* **2017**, *140*, e20170194. [CrossRef]

46. Austrian Agency for Health and Food Safety (AGES). Österreichische Agentur für Gesundheit und Ernährungssicherheit GmbH. Available online: https://www.ages.at/en/human/product-warnings-product-recalls (accessed on 10 March 2022).

47. AFSCA. Pour Les Consommateurs Agence Fédérale Pour La Sécurité De La Chaîne Alimentaire. Federal Agency for the Safety of the Food Chain. Available online: https://www.favv-afsca.be/consommateurs/avertissements/ (accessed on 10 March 2022).

48. E-Citizens Information and Service. HRana—Food-Related Warnings. Government of the Republic of Croatia. Available online: https://gov.hr/en/hrana-food-related-warnings/1774 (accessed on 10 March 2022).

49. Le Site Des Alertes De Produits Dangereux. Ministry of Agriculture and Food. France Government. Available online: https://rappel.conso.gouv.fr/categorie/ (accessed on 10 March 2022).

50. Lebensmittelwarnung. Das Bundesamt für Verbraucherschutz und Lebensmittelsicherheit, BVLD; Deutschland. Available online: www.lebensmittelwarnung.de (accessed on 10 March 2022).

51. Recent News. Hellenic Food Authority (EFET). Hellenic Food Authority. Available online: https://www.efet.gr/index.php/en/ (accessed on 10 March 2022).

52. Latest Food Alerts. Food Safety Authority of Ireland. Available online: https://www.fsai.ie/ (accessed on 10 March 2022).

53. Avvisi Di Sicurezza. Sicurezza Alimentare. Ministero della Salute. Governo Italiano. Available online: https://www.salute.gov.it/portale/news/p3_2_1_3.jsp?lingua=italiano&menu=notizie&p=avvisi (accessed on 10 March 2022).

54. Alertes Alimentaires. Sécurité Aimentaire. Le Gouvernement du Gran-Duché de Luxembourg. Available online: https://securite-alimentaire.public.lu/fr.html (accessed on 10 March 2022).

55. Food Alerts. Latest News. Environmental Health Directorate, Government of Malta. Available online: https://deputyprimeminister.gov.mt/en/environmental/Pages/Home-Page.aspx (accessed on 10 March 2022).
56. Veiligheidswaarschuwingen. Nederlandse Voedsel- en Warenautoriteit, (NVWA). Available online: https://www.nvwa.nl/onderwerpen/veiligheidswaarschuwingen/overzicht-veiligheidswaarschuwingen (accessed on 10 March 2022).
57. Rechemare/Retragere Produse Alimentare. Autoritatea Națională Sanitară Veterinară și pentru Siguranța Alimentelor, (ANSVSA); Guvernul României. Available online: http://www.ansvsa.ro/informatii-pentru-public/produse-rechemateretrase/ (accessed on 10 March 2022).
58. Obvestila Za Potrošnike. Uprava za varno hrano, veterinarstvo in varstvo rastlin.; Republika Slovenija. Available online: https://www.gov.si/drzavni-organi/organi-v-sestavi/uprava-za-varno-hrano-veterinarstvo-in-varstvo-rastlin/ (accessed on 10 March 2022).
59. Alertas Alimentarias De Alérgenos. Seguridad Alimentaria.Agencia Española de Seguridad Alimentaria y Nutrición. Ministerio de Consumo. Gobierno de España. Available online: https://www.aesan.gob.es/AECOSAN/web/subhomes/seguridad_alimentaria/aecosan_seguridad_alimentaria.htm (accessed on 10 March 2022).
60. HRana—Food-Related Warnings. Ministry of Agriculture. Government of the Republic of Croatia. Available online: https://hrana.mps.hr/HRana/ (accessed on 10 March 2022).

nutrients

Review

Food Allergy Education and Management in Schools: A Scoping Review on Current Practices and Gaps

Mae Jhelene L. Santos [1,2], Kaitlyn A. Merrill [2,3], Jennifer D. Gerdts [4], Moshe Ben-Shoshan [5] and Jennifer L. P. Protudjer [1,2,6,7,8,*]

1 Department of Food and Human Nutritional Sciences, Faculty of Agriculture, University of Manitoba, Winnipeg, MB R3T 2N2, Canada; santosmj@myumanitoba.ca
2 The Children's Hospital Research Institute of Manitoba, Winnipeg, MB R3E 3P4, Canada; kmerrill@chrim.ca
3 Department of Biochemistry, Faculty of Science, University of Winnipeg, Winnipeg, MB R3B 2E9, Canada
4 Food Allergy Canada, Toronto, ON M2J 4A2, Canada; jgerdts@foodallergycanada.ca
5 Division of Pediatric Allergy and Clinical Immunology, Department of Pediatrics, McGill University Health Centre, Montreal, QC H4A 3J1, Canada; moshe.ben-shoshan@mcgill.ca
6 George and Fay Yee Centre for Healthcare Innovation, Winnipeg, MB R3E 0T6, Canada
7 Department of Pediatrics and Child Health, Rady Faculty of Health Sciences, College of Medicine, University of Manitoba, Winnipeg, MB R3E 0W2, Canada
8 The Center for Allergy Research Karolinska Institutet, 171 77 Stockholm, Sweden
* Correspondence: Jennifer.Protudjer@umanitoba.ca; Tel.: +1-204-480-1384

Abstract: Currently, no synthesis of in-school policies, practices and teachers and school staff's food allergy-related knowledge exists. We aimed to conduct a scoping review on in-school food allergy management, and perceived gaps or barriers in these systems. We conducted a PRISMA-ScR-guided search for eligible English or French language articles from North America, Europe, or Australia published in OVID-MedLine, Scopus, and PsycINFO databases. Two reviewers screened 2010 articles' titles/abstracts, with 77 full-text screened. Reviewers differed by language. Results were reported descriptively and thematically. We included 12 studies. Among teachers and school staff, food allergy experiences, training, and knowledge varied widely. Food allergy experience was reported in 10/12 studies (83.4%); 20.0–88.0% had received previous training (4/10 studies; 40.0%) and 43.0–72.2% never had training (2/10 studies; 20.0%). In-school policies including epinephrine auto-injector (EAI) and emergency anaphylaxis plans (EAP) were described in 5/12 studies (41.7%). Educational interventions (8/12 studies; 66.7%) increased participants' knowledge, attitudes, beliefs, and confidence to manage food allergy and anaphylaxis vs. baseline. Teachers and school staff have more food allergy-related experiences than training and knowledge to manage emergencies. Mandatory, standardized training including EAI use and evaluation, and the provision of available EAI and EAPs may increase school staff emergency preparedness.

Keywords: anaphylaxis; epinephrine; food allergy; schools; scoping review; teachers

Citation: Santos, M.J.L.; Merrill, K.A.; Gerdts, J.D.; Ben-Shoshan, M.; Protudjer, J.L.P. Food Allergy Education and Management in Schools: A Scoping Review on Current Practices and Gaps. *Nutrients* **2022**, *14*, 732. https://doi.org/10.3390/nu14040732

Academic Editor: Carla Mastrorilli

Received: 11 January 2022
Accepted: 31 January 2022
Published: 9 February 2022

Publisher's Note: MDPI stays neutral with regard to jurisdictional claims in published maps and institutional affiliations.

1. Introduction

Food allergy affects an estimated 7.0–8.0% of children worldwide, or about two children in an average-sized classroom of 25 children [1–5]. A food allergy is defined by Boyce et al. (2010) as "a potentially life-threatening immunological response that occurs reproducibly upon ingestion of the allergen" (p. 11) and has the potential to result in severe allergic reactions [6]. Anaphylaxis, the most severe type of allergic reactions, was operationalized by Sampson et al. (2006) as a "potentially fatal condition that involves multiple organ systems or, when exposed to a known allergen, low blood pressure" [7]. Anaphylaxis affects an estimated 2.0% of the North American population [6], with similar estimates (between 0.3% [8] to 3.1%) noted in European populations [9,10].

Prior to the coronavirus disease (2019-nCoV/COVID-19) pandemic, about 20.0% of anaphylactic reactions occurred in schools [11–13], an observation that is unsurprising given that children typically spend the majority of their waking hours at school. Most in-school reactions occurred in the classroom, cafeteria, and playground [13–16]. Of concern is that an estimated 30.0% of allergic reactions occurred among children who were not previously known to have a food allergy or had an allergy that was not communicated to school staff [13,16].

Currently, policies surrounding food allergy management and its implementation are diverse both across and within jurisdictions [17–20]. Recently, international recommendations on the prevention and management for childcare centers and schools [11] was published based on the available scientific literature. Authors noted the utility of the guidelines as "conditional", wherein policymakers and stakeholders are to deliberate and adapt recommendations as needed to fit specific jurisdictional needs. Some of eight listed recommendations included school staff education and training, the removal of site-wide food bans and allergen-free zones, the requirement that children with a known food allergy had a current emergency anaphylaxis plan (EAP), and the availability of unassigned, or stock, epinephrine auto-injectors (EAI) in schools. Despite the need for further research in the topics described, this guideline may prompt jurisdictions to review and modify current policies.

The availability of EAI in school settings has been inconsistent. Students' access to and carriage of prescribed EAI also varies [21], and by socioeconomic advantage [22]. Even when a student has an EAI, school policy may render access difficult if it is locked in an office or exclusively carried by a staff member [12,13,16,21]. In cases where a prescribed EAI was unavailable, almost half of students requiring emergency medication were treated with stock epinephrine [23,24]. Additionally, trained staff available to administer EAI are also diverse. When available, school nurses administer EAI [13,14,23,25,26]. That said, only 50.0% of nurses reported food allergy management training, of which 35.0% described being "self-taught" [26]. As school nurses may work part-time [21] and among several schools [27], distributed responsibility and training among other school staff who are at school premises at all times is warranted. In brief, policies addressing stock EAI and EAP implementation are underused despite key recommendations and available resources [10,11,27,28].

Despite the above-described variation in policy, management, and treatment, there is, to our knowledge, no previous synthesis of the extant literature on teachers and school staff's knowledge and management practices of food allergy and anaphylaxis in schools. To this end, we aimed to conduct a scoping review on the in-school management of food allergies, and the perceived gaps or barriers in these management practices.

2. Materials and Methods

We performed a scoping review guided by the Preferred Reporting Items for Systematic Reviews and Meta-Analyses extension for Scoping Reviews (PRISMA-ScR) 2020 Checklist [29]. A literature search of original articles published in at least one of three medical literature databases (OVID-MedLine, Scopus, PsycINFO) was conducted on February 19, 2021. Search terms (see Supplementary Table S1) were identified in collaboration with content and methodological experts. Each search was filtered to child population and studies conducted in Canada, United States of America (USA), Australia, and Europe (including Turkey). Articles searched were restricted to publishing year 2006 and later to accommodate articles released subsequent to the implementation of Sabrina's Law, a law passed in 2006 following the fatal anaphylactic reaction of 13-year-old Sabrina Shannon, in a school in Ontario, Canada. Sabrina's Law requires every Ontario public school to implement an EAP for every student with food allergy including EAI administration instructions for staff [18].

Our primary outcome of interests were teacher and school staff management of food allergies in schools, including previous experience, knowledge and management of food allergy and anaphylaxis, emergency preparedness including availability of EAI and EAP, and school-based policies/guidelines. Studies were restricted to English and French. Additional inclusion criteria included previous experience in food allergy training, and experience working with students with food allergies, current practices, and food allergy knowledge of other school staff. There were no restrictions on type of study design. We excluded articles from grey literature, as well as abstracts, and publications without original data.

The search yielded 2010 articles (PsycInfo $n = 61$; Scopus $n = 1414$; OVID-MedLine $n = 535$). After the initial search and de-duplication (via Zotero $n = 299$; via Rayyan software [30] $n = 10$), there were 1701 articles, which were screened for titles and abstract by two independent reviewers (initials blinded for review; Figure 1). Titles/abstracts deemed potentially eligible for inclusion were advanced to full-text screening ($n = 77$). Full-text screening was made with consideration to study methods, participants, outcomes of interest, and findings. Full-text screening of English-language articles ($n = 75$) was conducted by two independent reviewers (initials blinded for review). French-language articles ($n = 2$) were full-text screened by a single reviewer (initials blinded for review) and excluded from the review. Two articles were reviewed by a third screener and were later excluded from the review [31,32].

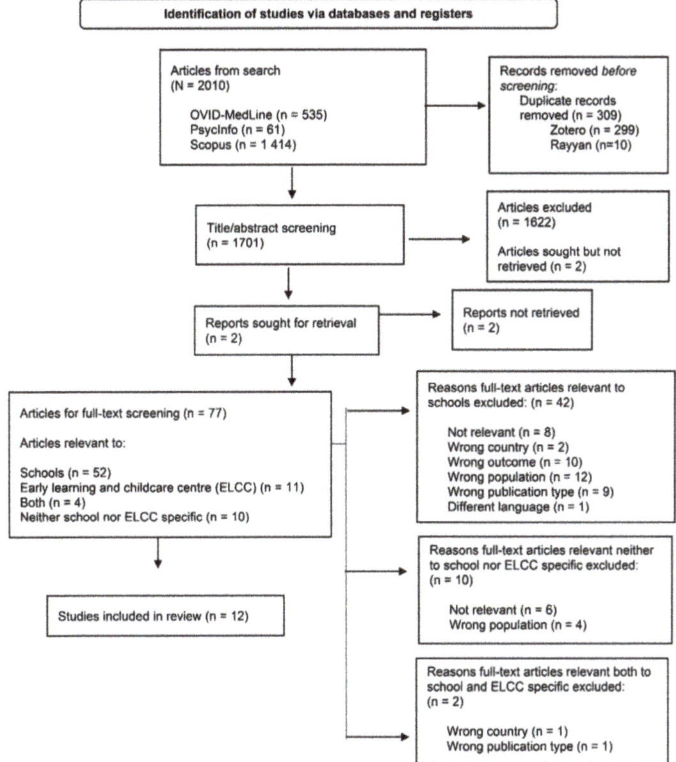

Figure 1. PRISMA flow diagram depicting the selection process articles and reports in the current scoping review.

As childcare centers may be housed in or proximate to schools, early learning and childcare centers were included in the initial search strategy. In the search strategy (Table S1), childcare centers were termed "daycare" and "daycare centers" and "preschool" as per recommendations from the expert librarian. However, owing to the developmental differences of children in schools vs. childcare centers, we restricted the present review to schools only and thereafter excluded studies that had aggregate data on school and childcare centers' teachers and staff. Data related to childcare centers will be reported elsewhere.

3. Results

From our search, 12 articles were included in the review, of which four (33.3%) studies were from North America [33–36] and eight (66.7%) [37–44] from Europe. About half of the studies (41.7%; 5/12) reported on teachers and school staff exclusively from primary school settings [34,37–39,42], 4/12 (33.3%) reported on mixed grade levels, the majority of which were primary schools [35,36,43,44], and 3/12 (25.0%) were presumed to represent primary schools [33,40,41] due to the language used, commonly differentiated in similar literature (e.g., "teachers" vs. "early childhood educators") (Table 1). Most included studies did not have, or did not specify, any school food program participation ($n = 10$), or school nurse availability ($n = 6$). Two studies (16.7%) reported that its schools had school food programs [34,36], while four (33.3%) studies reported that some participating schools had a part-time nurse [34–36,39], and two (16.7%) studies reported that the Italian public school system had no school nurses available [43,44].

Table 1. Summary of articles' country of origin, research design, methods, and population, presented in alphabetical order by first author's last name.

First Author, Year	Country	Research Design	Methods	Teachers and School Staff (n)	Type of School (n)
Polloni 2013 [43]	Italy	Quasi experimental pre/post-intervention	School staff attended an educational course by the Veneto Food Allergy Center and completed pre/post surveys.	1184 Teachers and Principals	Primary school ($n = 598$) Middle and high school ($n = 291$)
Polloni 2020 [44]	Italy	Quasi experimental pre/post-intervention	Teachers and school caretakers (class assistants and meal supervisors) participated in an educational intervention by the Veneto Food Allergy Center. The SPSMFAA questionnaire [32] was completed pre/post-session.	592 Teachers ($n = 474$) Caretakers ($n = 118$)	Primary school ($n = 216$) Middle and high school ($n = 152$)
Ravarotto 2014 [42]	Italy	Mixed methods (Focus group, pre/post-intervention)	Phase 1: 3–90-minute focus groups of teachers informed the intervention's communication strategy. Phase 2: Information workshop and *"The Theatre of Health"* show was held in various provinces. Phase 3: Teachers who attended the session completed pre/post questionnaires.	Three focus groups ($n = 25$ participants) Information workshop ($n = 197$) Assessment questionnaires ($n = 158$)	All primary schools. Focus groups ($n = 3$) Information workshops and questionnaire ($n = 5$)
Gonzalez-Mancebo 2019 [41]	Spain	Quasi experimental pre/post-intervention	"Management of Food Allergy in Children and Adolescents in School Centers" conference participants were provided an education session and a pre/post SPSMFAA questionnaire [32]. Training efficacy results between cafeteria monitors and teachers were compared.	191 Cafeteria monitors ($n = 97$) Teachers ($n = 46$) Cooks ($n = 25$); Other professions ($n = 23$)	Number of primary schools not reported
Rodríguez Ferran 2020 [40]	Spain	Multi-center quasi experimental pre/post-intervention	Teachers and canteen staff from three schools, as requested by patients' family members, participated in an educational session and pre/post questionnaire. Grade-specific data were not disclosed.	53 Teachers ($n = 45$) Canteen staff ($n = 8$)	Varied types of schools included. ($n = 3$) Schools had students aged 3–12y.

Table 1. *Cont.*

First Author, Year	Country	Research Design	Methods	Teachers and School Staff (*n*)	Type of School (*n*)
Ercan 2012 [39]	Turkey	Cross-sectional survey	Private and public-school teachers completed questionnaires, and food allergy knowledge was compared.	237 Public school teachers (*n* = 91) Private school teachers (*n* = 146)	Number of primary schools not reported
Ozturk Haney 2019 [38]	Turkey	Cross-sectional survey	Private and public-school teachers participated and completed the SPSMFAA questionnaire [32].	282 Public school teachers (*n* = 169) Private school teachers (*n* = 113)	All primary school (*n* = 12), of which 4 were private and 8 were public.
Canon 2019 [33]	USA	Multi-center pre/post-randomized intervention	Six Houston private schools were assigned to intervention (*n* = 4) or control groups (*n* = 2). Both groups completed the Chicago Food Allergy Research Survey [45]. Intervention groups received education sessions while control groups did not, and food allergy knowledge was compared.	375 Intervention (*n* = 302) Control (*n* = 73)	All private schools (*n* = 6)
Eldredge 2014 [36]	USA	Cross-sectional survey	Private, parochial schools participated in the survey. Electronic questionnaires were answered by principals or administrators. Grade-specific data were not disclosed.	78 Principals (*n* = 70) Administrators (*n* = 8)	Varied types of schools included. (*n* = 71) 76.0% were pre-K/K-6th or 8th grade.
Shah 2013 [34]	USA	Multi-center pre/post-randomized intervention	One school each from higher/lower socioeconomic areas in the Houston area were recruited. Intervention groups received education sessions while control groups did not, and food allergy knowledge was compared.	Pre-intervention (*n* = 195) Post-intervention (*n* = 131)	All public primary schools (*n* = 4)
Wahl 2015 [35]	USA	Quasi experimental pre/post-intervention	A school and community personnel training program provided education sessions and a survey. A follow-up survey was given 3–12-months post-intervention. Participants who participated in a food allergy emergency post-intervention were followed-up via phone interviews.	Primary survey (*n* = 4088) Secondary survey (*n* = 332) Phone interview (*n* = 21) Participant roles: Teachers (48%) Childcare providers (6%) School Aide (5%) Administrator (5%) School Nurses (2%) Other (34%) (Included camp counsellors, bus drivers, multiple of specified job titles, parents, volunteers, coaches, food service workers or no indication of job title)	Varied types of schools included. Number of primary schools not reported.
Raptis 2020 [37]	UK	Cross-sectional survey	All schools in the region were invited to participate in the survey. Only primary school data was presented in this study.	Specific participant roles not reported.	Primary schools (*n* = 157) High schools (*n* = 22) *

Abbreviations: *EAI* = epinephrine auto-injector; *K* = Kindergarten; *NS* = not specified; *SPSMFAA* = School Personnel's Self-efficacy in Managing Food Allergy and Anaphylaxis; *UK* = United Kingdom; *USA* = United States of America; *y* = years. * High school data were excluded in the paper per author reports.

Overall, food allergy experience, training and education, baseline knowledge, and policies/guidelines supporting food allergy management in schools were inconsistent between teachers and school staff, among and across jurisdictions.

3.1. Previous Experience in Food Allergy Management

The majority of teachers and school staff had experience with food allergies, as reported in 10/12 (83.3%) studies. However, higher proportions of teachers and school staff reported caring for a child with a food allergy compared to the teachers and school staff who had received training to do so.

An estimated 20.0–88.0% of Turkish, Italian, English, and American teachers and school staff reported having students with food allergies [33,37,38,42]. One study reported that 44.7% of Italian teachers had 1–2 students with a food allergy in their teaching experience, 31.6% had 3–5 students, and 23.7% had >5 students [42]. On average, United Kingdom (UK) schools enrolled between 1–12 students with a food allergy per school [37]. One Turkish study reported only 53.2% of participating teachers knew which students had a food allergy [39]. Fewer teachers (3.0–9.0%) reported they had taught students with a history of anaphylaxis than a food allergy [38,40]. Among UK schools, 57.0% (n = 89/157) reported having students who had previously had severe allergic reactions [37].

Rates of prior food allergy education were variable. Among Italian, Turkish, and Spanish teachers and school staff, rates of food allergy training ranged from 14.0–63.6% [38,41–44], whereas 43.0–72.2% of Italian [42] and Spanish [41] teachers and school staff reported no previous food allergy training at all. The majority of Italian and Turkish teachers (71.7–82.3%) reported having first aid training, although the extent of food allergy training included (e.g., EAI administration) was unspecified [38,43,44]. In Washington state, USA, approximately half (51.1%; 1102/2156) of teachers reported previous food allergy training. Of these same teachers, 62 reported having administered an EAI, although not all (77.4%; 48/62) had prior EAI training [35].

The method of food allergy education delivery was reported in 40.0% (4/10) of studies. Italian, Turkish, and American teachers and various school staff received previous food allergy education primarily from first aid courses (71.7%) [34,43], health training (11.1%) [43], mass media (22.4–64.5%) [39,43], the internet (17.9–23.0%) [39,43], booklets (37.3%) [39], seminars (22.4%) [39], and less commonly, via acquaintances or relatives (1.4%) [43]. Other sources of food allergy information included sessions from in-service days and/or regional conferences [39], parents, and individuals with a food allergy [42].

3.2. Baseline Knowledge

Teachers and school staff reported poor knowledge of food allergy understanding and anaphylaxis management at baseline in 6/12 (50.0%) studies from Italy, Spain, USA, and Turkey [34,38–40,42,43]. Turkish and Italian teachers and school staff had knowledge of allergic reaction symptoms, but a poor understanding of food allergy and anaphylaxis management [39,42,43]. Notably, Italian teachers and principals from primary schools had statistically significant higher baseline questionnaire scores than middle schools (p < 0.001) when compared through one way analysis of variance and Bonferroni post hoc test [43].

The majority of American primary school teachers from economically-advantaged and disadvantaged areas (78.3% and 76.5%, respectively) [34] and one group of Italian teachers and school staff of various grade levels (79.3%) were able to identify common allergenic foods [43], compared to approximately 40.0% of Turkish primary school teachers, and another group of Italian primary school teachers, who correctly answered questions about food as allergic triggers [39,42]. Interestingly, one group of primary school Italian teachers acknowledged having poor food allergy knowledge (mean = 5.1/10; Standard Deviation (SD) = 2.1) but perceived food allergy as a significant issue in schools (mean = 7.6/10, SD = 2.1, based on a scale of 1–10, with higher scores corresponding to higher significance) [42].

The economic advantage of school areas appeared to also influence teacher and school staff's baseline food allergy knowledge. Primary school teachers in both Houston, USA, and Turkey from schools in economically-advantaged areas had more non-statistically significant higher baseline food allergy knowledge than teachers from economically-disadvantaged areas [34,39].

Anaphylaxis knowledge was likewise poor as reported in 3/12 (25.0%) of studies from Italy, Spain, and the USA [34,40,43]. Italian and Spanish authors determined that teachers' and school staff's baseline knowledge was not influenced by previous education on food allergy or experience working with students with an anaphylaxis history [40,43].

An estimated 45.3% of Spanish primary school teachers [40] and 65.4% of Italian [43] teachers of various grade levels correctly reported that epinephrine is the main anaphylaxis treatment. Similar rates of anaphylaxis treatment knowledge were reported by American teachers from economically-disadvantaged areas (45.3–49.0%) compared to teachers from economically-advantaged areas (70.0–80.6%) [34]. Conversely, fewer Italian teachers and principals of various grade levels (34.5%) knew epinephrine was safe to use for suspected anaphylaxis without severe side effects [43]. Fewer Spanish and Turkish primary school teachers and canteen staff knew what an EAI was (10.1% [39], and 18.9% [40], respectively), or how to use an EAI (6.8–13.2%) [39,40] and where to administer it (3.8%) [39]. If faced with a food allergy-related emergency, only 24.5% of Turkish primary school teachers stated they would administer first aid, although none of the teachers identified that epinephrine was the appropriate medication to use [39].

3.3. In-School Emergency Preparedness

Food allergy-related emergency preparedness, with regard to self-efficacy, confidence, and food allergy-related emotions, was discussed in 6/12 (50.0%) of studies, all of which were European [37,38,41–44].

Self-efficacy in managing food allergies in school was discussed in three studies, all of which made use of the School Personnel's Self-Efficacy in Managing Food Allergy and Anaphylaxis (SPSMFAA) questionnaire by Polloni et al. (2016) [32] to measure self-efficacy on food allergy management. The questionnaire measures a total of 40 points based on eight factors (1 = cannot do, 5 = highly certain can do) [32]. Compared to anaphylaxis management, food allergy management was associated with greater self-efficacy [38,41,44]. Turkish primary school teachers exhibited that previous food allergy experience and food allergy training were associated with greater self-efficacy in managing a food allergy and anaphylaxis ($p < 0.001$) [38]. In fact, significant SPSMFAA score differences were seen among Turkish primary school teachers with previous food allergy training compared to those who did not have previous training (mean = 26.74/40 ± 6.21, vs. 22.18/40 ± 7.48, respectively; $p < 0.001$) [38].

Confidence in managing anaphylaxis was reported by approximately half (47.3%; 53/112) of UK primary schools, with no difference ($p = 0.10$) among schools with or without students with a food allergy (52.6% vs. 36.1%, respectively) [37]. Most UK schools (60.7%) demonstrated being prepared for allergic reactions in students without a previous allergic history by establishing communication and documentation systems, and identifying staff member roles in the event of an allergic emergency, with no significant difference between schools with vs. without students with food allergy enrolled (61.0% vs. 60.0%, respectively; $p = 0.94$) [37].

Elsewhere, Italian teachers and principals of various grades reported food allergy-related emotions were concern (66.9%), anxiety (15.8%), fear (3.7%), and helplessness (7.0%). Positive attitudes were also associated (9.3%) in relation to newfound post-intervention knowledge [43].

Three focus groups of Italian primary school teachers ($n = 25$) qualitatively discussed concerns over managing the child in crisis and other students in class [42]. Teachers were unauthorized to administer certain (unspecified) drugs, thus, had restricted emergency management abilities to providing first aid and calling for help. It was not disclosed what type of first aid treatment teachers were allowed to perform. Feelings of insecurity were described, and teachers felt unable to manage emergencies due to the perceived lack of food allergy knowledge. Additionally, teachers thought that the responsibility of food allergy management was beyond their teaching duties and required more emotional involvement [42].

3.4. School-Based Policies and Guidelines

School-based policies/guidelines were described in 5/12 (41.7%) of studies, although implementation and adherence were variably enforced among participating schools [34,36,37,39,40]. An outline of policies and guidelines are listed in Table 2.

Table 2. Summary of in-school policies, emergency action plan, epinephrine auto-injector availability, and other management practices among schools, presented in alphabetical order by first author's last name.

First Author, Year	Policies	EAP Availability	EAI Availability	Other Management Practices
Eldredge 2014 [36]	71.0% of schools had some sort of guideline/policy for food allergy while 25.0% of schools had none.	56.0% of schools required an EAP.	Not reported	76.0% of schools needed special arrangements (i.e., peanut-free classroom, allergen-free areas or cafeteria tables, increased monitoring, physical distancing, and having special meals for students with food allergy). 57.0% of schools had handwashing guidelines. 30.0% had no food sharing policies. 58.0% had classroom project food substitution guidelines and 45.0% had cleaning surfaces with allergen contact.
Ercan 2012 [39]	Not reported	6.0% of teachers, all from private schools, had available EAP. 86.0% of teachers had no EAP, and 8.0% were uncertain if EAPs were available.	Not reported	Not reported
Raptis 2020 [37]	76.0% of schools had standard protocols related to allergic reactions.	89.5% of schools reported having an EAP for students with anaphylaxis history.	0.7% (n = 165) of students with food allergy had prescribed EAI. 45.2% of schools reported their students at risk of anaphylaxis carried an EAI.	Schools had guidelines for: staff food handling guidelines (79.0%), special mealtime supervision (49.0%), no food sharing policy (63.0%), no utensil sharing policy (45.0%), aware of food packaging regulations (66.0%), reviewed curriculum to remove allergen foods (68.0%), and no eating on transportation policy (48.0%), communication systems during emergencies (94.1%), identifying staff roles (82.1%), documenting staff emergency response (81.9%), and preparing for allergic reactions in students without prior allergic history (60.7%).
Rodriguez Ferran 2020 [40]	Not reported	83.0% of teachers and school staff reported they had EAP.	66.0% of teachers and school staff knew where EAI was in their school.	56.0% of teachers and school staff had meetings with parents/guardians of students with food allergy in their care.
Shah 2013 [34]	Not reported	Not reported	Schools in economically-disadvantaged areas had 1 EAI each. Schools in economically-advantaged areas had 6 and 9 EAI each.	Not reported

Abbreviations: *EAI* = epinephrine auto-injector; *K* = Kindergarten; *NS* = not specified; *SPSMFAA* = School Personnel's Self-efficacy in Managing Food Allergy and Anaphylaxis; *UK* = United Kingdom; *USA* = United States of America; *y* = years.

EAP usage was inconsistently implemented (5.9–89.5%) among participating schools from Italy, UK, and the USA [36,37,39,40]. EAI was available, as prescribed in one UK study [37,42], and unspecified in one Spanish study [40]. In Spanish schools where EAI was available, only 66.0% of teachers and school staff reported to know where it was located [40]. One Houston, USA-based study stated more stock EAI was available in two schools in economically-advantaged areas (n = 6–9 per school) compared to two schools in economically-disadvantaged areas (n = 1 each) [34].

Food bans and mealtime accommodations were the most common policies imposed in schools as reported by 3/12 (25.0%) of the Milwaukee, USA; Spanish; UK studies [36,37,40]. Other preventative policies implemented among these schools were distancing measures, e.g., separate lunch table for students with food allergies, safe food/utensil handling, handwashing, surface cleaning, food sharing, and reviewed food items for classroom projects [36,37]. Teachers were primarily responsible for carrying out tasks to manage food allergies such as mealtime supervision [36,37] and meeting with the parents and students with food allergies [40].

In a study by Eldredge et al. (2014), of which 76.1% of responding Milwaukee schools included primary school students, the authors reported on rates of food allergy policy implementation. Authors also noted that policies in this school district were independently determined by governing parishes and/or school boards. Nevertheless, enrollment of students with food allergy appeared to determine policy/ guideline implementation. In this study, 71.0% (53/75) of schools reported some policy/guideline in place. Schools with students with food allergies had an increased likelihood of implementing policies compared to schools without students with a food allergy (Odds Ratio (OR) = 6.30, 1.50–2.60). In fact, 85.0% of schools who had students with a food allergy enrolled had policies implemented, compared to the 15.0% of schools without policies ($p \leq 0.0001$). Schools with policies were also 3.5 times more likely to require EAPs than schools without policies (67.0% vs. 35.0%, respectively; $p < 0.0001$; OR = 3.50, 95% Confidence Interval (CI) = 1.00–12.20) [36].

In a UK study of primary schools, 76.0% (111/152; 95% CI = 68.0–83.0%) reported having a standard management protocol. An estimated 0.7% (165/24,174) of students had a history of anaphylaxis, or were at risk for severe reactions, and had an EAI. Compared to schools at which there were no students at risk for anaphylaxis, schools attended by students at risk were significantly more likely to have a standard management protocol (57.0% vs. 90.0%, respectively; $p < 0.001$) [37].

3.5. Post-Educational Intervention Knowledge

Interventional education sessions were described in 8/12 (66.7%) of studies. Sessions were delivered through a healthcare provider-led presentation. One-third (4/12; 33.3%) of studies also provided hands-on EAI training [35,40,41,44].

Overall, teachers and school staff who received interventional education demonstrated better knowledge on food allergy and anaphylaxis management [33–35,40–44] compared to their baseline knowledge or versus controls [33,34]. The key outcomes of each study are listed in Table 3.

Table 3. Summary of studies that provided educational interventions (n = 8), presented in alphabetical order by first author's last name.

First Author, Year, Country	Intervention and Session Topics	Key Intervention Outcomes
Canon 2019 [33] USA	1-hour education session with HCP Sessions taught case scenarios, common food allergens, routes of exposure, reaction recognition and prevention, epinephrine administration, importance of EAP, bullying of students with food allergy and classroom protocols.	Intervention group had higher post-intervention survey scores compared to controls (95% CI = 16.62–22.53; $p < 0.001$) and their pre-test surveys (95% CI = 18.17–21.38; $p < 0.001$). Intervention vs control group post-intervention were more likely to recognizing food allergy as life-threatening and agree that children with food allergy were treated differently and bullied ($p < 0.001$), 5 times more likely to acknowledge food avoidance is hard ($p = 0.003$) and 874 times more likely to agree that EAI is an important lifesaving measure and use it in an emergency ($p = 0.173$).

Table 3. *Cont.*

First Author, Year, Country	Intervention and Session Topics	Key Intervention Outcomes
Gonzalez-Mancebo 2019 [41] Spain	Education session and EAI workshop for school staff included practical EAI training. Sessions taught food allergy definition, diagnosis, problems of children with food allergy in school settings, allergic reaction recognition and prevention measures, coordination of care, anaphylaxis treatment and, and EAP discussion	Significant improvements in SPSMFAA questionnaire [32] mean scores were observed ($p < 0.05$). The largest pre-post mean score difference was in managing allergen avoidance (mean = 4.29, SD = 0.98 vs. mean = 4.51, SD = 0.72). The smallest difference was in administering drugs (e.g., EAI) to a student having a severe and sudden reaction (mean = 3.08, SD = 1.41 vs. mean = 4.51, SD = 0.84). Case study scores also improved from pre- post intervention (25.5% vs 96.9%, respectively).
Polloni 2013 [43] Italy	2-hour session with a pediatric allergist, dietician, psychologist, and a lawyer. Session topics were not specified.	Primary school teachers scored higher than nursery or high schools (F-value: 13.450, df = 2, $p < 0.001$). Mean scores significantly increased from pre-post-intervention. From pre-post-intervention, more participants thought anaphylaxis could be managed in schools (82.6% vs. 96.5%, respectively; $p < 0.001$) and is school staff responsibility (82.8% vs. 93.9%, respectively; $p < 0.001$). Feelings related to food allergy management were concern (66.9%), anxiety (15.8%), fear (3.7%) and helplessness (7.0%).
Polloni 2020 [44] Italy	2-hour session with an allergist, psychologist, and a lawyer. Practical EAI training was included. Sessions taught description of allergic mechanisms, signs and symptoms, prevention and treatment, explanation of EAPs and presentation of national and regional regulations on food allergy-related drug administration in schools and discussions on food allergy-related psychosocial and emotional issues.	Improvements in SPSMFAA questionnaire [32] mean scores were observed. Post–pre score differences in anaphylaxis management (0.67–1.67, respectively), was higher than food allergy management difference (0.2–1.0, respectively). The largest pre-post mean SPSMFAA [32] score difference was in administering drugs (e.g., EAI) to a student having a severe and sudden reaction (mean = 1.3) and the lowest in guaranteeing students with food allergy full participation to all school activities (mean = 0.47). Median scores increased, as evaluated through conditional regression, from pre-post-intervention (<17 to 25, respectively), independent of all other covariates (type of job, age, school, gender, previous anaphylaxis and food allergy knowledge, training, and experience).
Ravarotto 2014 [42] Italy	2-hour workshop with allergist or pediatrician, a veterinarian, and a scientific communication expert. Sessions taught common allergenic foods, difference between allergy and intolerance, allergic reaction signs and symptoms, first aid introduction, available training tools/ resources and regulations to protect consumers	The number of correct answers determined knowledge categories. Pre-intervention, 3.2% had poor knowledge, 56.3% had fair, 39.9% had satisfactory, and 0.6% had good knowledge. Post-intervention, the percentage of correct answers increased to 1.3% fair, 67.7% satisfactory, and 31.0% good knowledge. Increased knowledge was unrelated to previous food allergy training ($\chi2 = 0.143$, $p = 0.931$).
Rodríguez 2020 [40] Spain	40–50-minute presentation by pediatric allergist and a 10–20-minute EAI practical session by pediatric nurse. Sessions taught allergy definition allergic reactions pathophysiology, reactions prevention and recognition, communication with family and EAP development, anaphylaxis management, legal aspects and official recommendations.	From pre-post-intervention, participants had significantly better anaphylaxis recognition (40.0% vs. 81.0%, respectively; $p < 0.001$). Knowledge of how and when to use the EAI increased from 19.0% and 13.0%, respectively, to 100.0% of participants ($p < 0.001$).
Shah 2013 [34] USA	1-hour education session with physician. Sessions taught food allergy prevalence, causal foods, signs of local and systemic reactions, reaction prevention and treatment.	Teachers in the economically-disadvantaged vs. economically-advantaged school areas had a larger increase in correct answers post-intervention (34.6%; 95% CI = 32.1–103.9 vs. 24.6%, 95% CI = 21.5–74.1, respectively). Teachers from both economically-disadvantaged and advantaged school areas had increased scores from pre-post-intervention in questions related to treatment of local and systemic reactions, causal foods, and signs of anaphylaxis.

Table 3. *Cont.*

First Author, Year, Country	Intervention and Session Topics	Key Intervention Outcomes
Wahl 2015 [35] USA	45-minute presentation by a food allergy Educator nurse. Practical EAI training was included. Sessions taught key food allergies facts, allergic reactions, prevention, and recognition, and importance of immediate treatment.	Post-intervention, most teachers and school staff had better confidence in prevention of allergic reactions (94.0%), recognizing reaction signs and symptoms (96%), know what to do in an emergency (97%), and administer an EAI (94%). Approximately half of participants had prior food allergy training. 95.0% of participants had positive feedback about food allergy management confidence in preventing allergic reactions, symptom recognition, and knowing what to do in emergencies 3–12-months post-intervention. 57.0% of participants recalled three key messages from the sessions. 21 participants who experienced a food allergy emergency post-intervention were interviewed. 61.9% found that signs and symptoms recognition and 52.3% reported EAI training were useful in real-life situations.

Abbreviations: *EAI* = epinephrine auto-injector; *HCP* = healthcare professional; *UK* = United Kingdom; *USA* = United States of America.

Sustained knowledge and confidence levels were also described in one American longitudinal study that followed-up with participants, including teachers and school staff from various grade levels, 3–12 months post-intervention. Participants reported sustained confidence levels in the recognition of signs and symptoms, ability to prevent food allergic reactions, and knowing what to do during an anaphylaxis emergency [35]. Primary key messages recalled by 57.0% of participants 3–12 months post-intervention included EAI administration, reaction signs and symptoms, importance of following an EAP, and providing immediate treatment [35]. A small proportion of participants ($n = 22$) experienced a food allergy emergency post-intervention, 42.8% of which were caused by unknown allergens and 23.8% occurred in primary schools. Of these participants, 81.8% (18/22) had previous training before the study intervention. Nevertheless, 61.9% found that the recognition of food allergic signs and symptoms and 52.3% found the hands-on EAI training useful in real-life situations [35].

In a Houston, USA-based study, the intervention group teachers from economically-disadvantaged school areas had non-significant higher questionnaire scores post-intervention than teachers from economically-advantaged schools in both intervention and control groups [34]. Another Houston study that compared teachers who received intervention to those who did not, reported that there was no correlation between level of education (<4 years college, 4 years college, and graduate degree) and the survey scores [33]. Spanish primary school teachers and school staff exhibited significantly better food allergy knowledge ($p < 0.001$) through improved recognition of anaphylaxis (40.0% to 81.0%, respectively), knowledge about when (19.0% vs. 100.0%, respectively) and how (13.0% vs. 100.0%, respectively) to use an EAI, albeit authors reported modifying acceptable questionnaire responses as the original questions were "not easy to answer" [40]. Education sessions were deemed useful by Italian primary school teachers ($8.6/10 \pm 1.67$; on a scale of 1–10, where 10 = very useful) [42]. Another group of Italian teachers and principals from various grade levels showed significantly better questionnaire scores post-intervention (mean = 6.6/10 vs. 8.9/10, respectively; $p < 0.001$) [43]. Post-education, the same Italian group of teachers and principals agreed anaphylaxis is manageable at school (82.6% vs. 96.5%, respectively; $p < 0.001$) and school staff are responsible for food allergy management (82.8% vs. 93.9%, respectively; $p < 0.001$) [43].

Interventional education influenced teachers and school staff's beliefs and attitudes about food allergy management. Among Houston, USA-based private school teachers, those in the intervention group, compared to control group teachers who did not receive intervention, tended to show greater agreement about the importance of EAI as a lifesaving measure for anaphylaxis. Although the authors identified an OR = 873.77 ($p = 0.173$), the difference was statistically insignificantly different because, as the authors noted, "almost all" participants agreed or strongly agreed with the importance of EAI [33]. Similarly, compared to the baseline, intervention group teachers were 3.3 times more likely to recognize the seriousness of food allergies (OR = 3.30; 95% CI = 1.60–6.70; p = 0.001) and to agree that students with food allergies are likely to experience discrimination (OR = 3.30; 95% CI = 2.00–5.50; p = 0.01) [33]. Intervention teachers were also 52 times (OR = 52.0; 95% CI = 2.90–930.75; p < 0.01) more aware, post-intervention, that students with food allergies experienced bullying compared to control teachers, with 26 times increased likelihood of agreement that students with food allergies experienced bullying (OR = 25.55; 95% CI = 9.86–66.25; p < 0.001) [33].

Education sessions were associated with increased confidence [35], comfort level [34], and self-efficacy [41,44] in the majority of participants, regardless of whether participants had previous training [35,41,42,44]. The majority of American participants (>94.0%), some of whom were teachers and school staff, answered opinion statements positively post-intervention, indicating more confidence in prevention, recognition, and response skills to food allergy emergencies [35]. Significant post-intervention SPSMFAA scores [32] were reported for Spanish teachers and school staff (p < 0.05) in food allergy management items, specifically in putting an EAP in place for students with a food allergy, managing students at risk of reactions to food, and recognizing anaphylaxis symptoms and administering EAI in anaphylaxis management [41]. Following a food allergy intervention, Italian teachers' and school staff's post-intervention scores were higher compared to pre-intervention studies. The greatest differences were seen among those with low self-efficacy at baseline [44].

3.6. Future Educational Needs

The majority of primary school teachers and staff expressed an interest in receiving more food allergy and anaphylaxis training [36,37,39,42]. Teachers also thought that increasing food allergy awareness in schools and involving all students may increase empathy among all schoolchildren [42]. To deliver further food allergy education and awareness, study participants suggested establishing online repositories for educational resources, have more in-person training or video training [36,42], and have students with food allergies wear medical alert accessories to inform others of their condition [39]. Additionally, nearly all (94.0%) of UK teachers either "agreed" or "strongly agreed" that unprescribed EAI ought to be kept in schools [37]. Interestingly, schools with no students at risk of anaphylaxis were non-statistically significantly more likely to agree than schools with students at risk of anaphylaxis (55.6% vs. 30.3%, respectively; p = 0.09) [37].

4. Discussion

In this scoping review of the European and North American literature on in-school management of food allergies, we identified several perceived gaps and barriers in management. First, teachers and school staff acknowledged the significance of food allergies [42] yet lacked experience and knowledge. We identified participants' knowledge differences [33,39] and EAI availability [34] from schools in economically-advantaged and disadvantaged areas. Studies also reported that teachers and school staff did not know which students had a food allergy [37,39]. Second, there exists wide variation, and reporting, of food allergy management practices including the provision of policies/guidelines, EAP implementation, and inconsistency in EAI availability and knowledge in EAI administration, as similarly described in other studies [13,14,22,24,25]. Third, preparedness and self-efficacy of teachers and school staff to manage anaphylaxis effectively are correspondingly variable. Unsurprisingly, additional training was desired by many.

The need for additional training is underscored by the commonality of students with food allergies, juxtaposed against inconsistent policies across and between jurisdictions [17–20]. As school staff are likely to be the first adults to be notified of food allergy-related emergencies [15], adequate and universal emergency management skills are essential in student safety, including EAI administration. One USA-based study in our review reported that not all teachers have administered EAI but have not been previously trained [35], which illustrated that teachers are key players in emergency management in schools, especially when there are no school nurses available. School nurses have also reported to have inconsistent training, or were "self-taught" [15,25,26]. Reliance on one nurse to manage medical emergencies may be impractical as allergic reactions can occur anywhere within school premises. Additionally, if parents are less involved and/or unaware of serious food allergy concerns, teachers may also assume caregiving responsibilities and help students learn about their own food allergy management.

Our review highlights the need to share food allergy management responsibilities, including, but not limited to, maintaining individual EAPs, knowing where EAI are located and how to use it, promoting preventative practices (e.g., handwashing) and recognizing signs and symptoms of allergic reactions, and knowing own roles in emergencies by providing food allergy training for all teachers and school staff, including school nurses where available. Such training may also reduce the propensity of other school staff to turn to online, non-academic resources for food allergy education [39,42,43]. Moreover, early (pre-hospital) treatment decreases the risk of hospitalization [13], while delayed treatment from symptom onset was associated with the risk of having a biphasic reaction and fatality [12,24]. As the long-term effects of staff training on food allergy management knowledge are unknown, the implementation of post-training evaluation may also be beneficial [11].

School meal programs also raise the value of food allergy training for other school staff such as cafeteria personnel and food monitors, as proper food handling and preparation are foundational in preventing allergic reactions [6,46]. Our study reported on two studies with school food program participation that did not discuss how food allergies were accommodated [34,36]. Future training programs should also address how school food programs apply food allergy education in practice, including safe food handling training, cleaning protocols, and increased mealtime supervision for younger students who may have more impulsive behaviors [47].

Although a universally accepted EAP and laws to provide stock epinephrine in schools would prove challenging to develop and garner acceptance, we purport that such calls are essential at a national, or regional level. Collaborative efforts and partnerships among all stakeholders including affected students and families should focus on identifying students at risk of anaphylaxis. Thus, planning and implementation of medically sound EAPs, yet relevant and clearly understood by its intended users, is essential. Additionally, in conjunction with staff training and the implementation of EAPs, stock EAI in schools would be advantageous as not all students with a food allergy may have an EAI, or do not carry them around school. Meetings with teachers, children, and their families may also increase communication and consensus on stock EAI usage and care plans [40]. Likewise, training, EAP implementation, and stock epinephrine availability align with international recommendations [11,28], and may increase staff awareness of food allergies, and help alleviate concern, anxiety, fear, and helplessness reported by teachers and school staff [43]. In turn, training may contribute to teachers and school staff's confidence, self-efficacy, knowledge, and ability to perform in emergency situations.

To our knowledge, this is the first scoping review to provide an overview in some school jurisdictions in Europe and USA. We did not restrict the publications to the English language only and presented available data from multiple Western countries. Our review also extends the findings from Waserman et al. (2021), such as the positive uptake and perceived benefits of teachers and school staff of food allergy training, providing available EAI and implementation of action plans [11].

We acknowledge that searching only within three databases and the publication year cut-off may have introduced some reporting bias and reduced eligible studies. We also did not perform a quality appraisal of the included studies or comparisons of the interventions. Moreover, our ability to compare the interventions and results into a cohesive analysis were limited given the heterogeneity of design of the included studies [48]. However, we were able to identify common themes. We recognize that we excluded all grey literature, as well as publications outside Europe and North America, and in languages other than English or French.

Nevertheless, our review highlights several key take-away messages (Box 1), including the need for further research and the creation of a food allergy training strategy that includes EAI administration for all school staff. Our review findings can also be used to inform policymakers to consider implementing an evaluation program for existing training courses. In light of the COVID-19 pandemic, the usage of virtual platforms for training purposes can be an accessible communication medium. Lastly, the provision of stock EAI and individualized EAPs should be considered as mandatory as jurisdictions are able. The execution of such may pose greater benefits beyond having available rescue medication but may also help increase the confidence and self-efficacy of teachers and staff to be able to manage emergency situations appropriately.

Box 1. Key take-away messages.

> - Teachers and school staff play a pivotal role in emergency response.
> - At baseline, teachers and school staff have poor and variable knowledge and experience of food allergy.
> - Teachers and school staff may benefit from standardized, annual food allergy training.
> - Key elements of food allergy training may include epinephrine auto-injector (EAI) administration, causal foods, signs and symptoms of a reaction, and importance and usage of a emergency anaphylaxis plans (EAP).
> - Implementation of EAP for all students with a food allergy and having stock EAI, in conjunction with annual training will improve student safety and schools' emergency preparedness.

5. Conclusions

In conclusion, current in-school management of food allergies, including food allergy education, are highly heterogeneous across jurisdictions in western nations for which data are available.

Implementation, continuation and/or evaluation of universal standardized training, usage of personalized EAPs, provision of stock EAI in schools, and policy or guideline implementation outlining these practices may be considered by schools and governing jurisdictions. As such, these actions will support teachers and staff in preventing and managing in-school food allergy emergencies safely and effectively.

Supplementary Materials: The following are available online at https://www.mdpi.com/article/10.3390/nu14040732/s1, Table S1. Search strategy.

Author Contributions: Conceptualization, J.L.P.P. and M.J.L.S.; Methodology, J.L.P.P. and M.J.L.S.; Data curation, M.J.L.S., K.A.M. and J.L.P.P., Writing—original draft preparation, M.J.L.S.; Writing—review and editing, M.J.L.S., K.A.M., J.L.P.P., M.B.-S. and J.D.G.; Supervision, J.L.P.P.; Funding acquisition, J.L.P.P. All authors have read and agreed to the published version of the manuscript.

Funding: M.J.L.S. receives funding for her MSc studies from a grant held by J.L.P.P. from Canadian Institutes of Health Research (Application number 433502), with matched funding from the Children's Hospital Research Foundation, and supported with additional funds for M.J.L.S. from the University of Manitoba Graduate Enhancement of Tri-Agency Stipends (GETS) Program.

Institutional Review Board Statement: Not applicable as this was a scoping review.

Informed Consent Statement: Not applicable as this was a scoping review.

Acknowledgments: Thank you to Mê-Linh Lê for guiding the search of this scoping review, and Natalie Riediger for her ongoing co-supervision on Mae Santos' master's thesis research.

Conflicts of Interest: M.J.L.S. declares no conflicts of interest. K.M. declares no conflicts of interest. J.D.G. is Executive Director of Food Allergy Canada, and co-leads Canada's National Food Allergy Action Plan. M.B.-S. is a Member of the Board, Canadian Society of Allergy and Clinical Immunology, and on the steering committee for Canada's National Food Allergy Action Plan. J.L.P.P. is Section Head, Allied Health and Member of the Board, Canadian Society of Allergy and Clinical Immunology; sits on the steering committee for Canada's National Food Allergy Action Plan; and reports consultancy for Novartis and ALK Abelló.

References

1. Clarke, A.E.; Elliott, S.J.; St Pierre, Y.; Soller, L.; La Vieille, S.; Ben-Shoshan, M. Temporal trends in prevalence of food allergy in Canada. *J. Allergy Clin. Immunol. Pract.* **2020**, *8*, 1428–1430.e5. [CrossRef] [PubMed]
2. Venkataraman, D.; Erlewyn-Lajeunesse, M.; Kurukulaaratchy, R.J.; Potter, S.; Roberts, G.; Matthews, S.; Arshad, S.H. Prevalence and longitudinal trends of food allergy during childhood and adolescence: Results of the Isle of Wight Birth Cohort study. *Clin. Exp. Allergy* **2018**, *48*, 394–402. [CrossRef] [PubMed]
3. Gupta, R.S.; Springston, E.E.; Warrier, M.R.; Smith, B.; Kumar, R.; Pongracic, J.; Holl, J.L. The Prevalence, Severity, and Distribution of Childhood Food Allergy in the United States. *Pediatrics* **2011**, *128*, e9–e17. [CrossRef] [PubMed]
4. Rona, R.; Keil, T.; Summers, C.; Gislason, D.; Zuidmeer-Jongejan, L.; Sodergren, E.; Sigurdardottir, S.T.; Lindner, T.; Goldhahn, K.; Dahlstrom, J.; et al. The prevalence of food allergy: A meta-analysis. *J. Allergy Clin. Immunol.* **2007**, *120*, 638–646. [CrossRef]
5. Fraser Institute. Available online: https://www.fraserinstitute.org/studies/secondary-school-class-sizes-and-student-performance-in-canada (accessed on 6 October 2021).
6. Boyce, J.A.; Assa'Ad, A.; Burks, A.W.; Jones, S.M.; Sampson, H.A.; Wood, R.A.; Plaut, M.; Cooper, S.F.; Fenton, M.J.; Arshad, S.H.; et al. Guidelines for the Diagnosis and Management of Food Allergy in the United States: Report of the NIAID-Sponsored Expert Panel. *J. Allergy Clin. Immunol.* **2010**, *126*, S1–S58. [CrossRef]
7. Sampson, H.A.; Muñoz-Furlong, A.; Campbell, R.; Adkinson, N.F.; Bock, S.A.; Branum, A.; Brown, S.; Camargo, C.A.; Cydulka, R.; Galli, S.J.; et al. Second symposium on the definition and management of anaphylaxis: Summary report—Second National Institute of Allergy and Infectious Disease/Food Allergy and Anaphylaxis Network symposium. *J. Allergy Clin. Immunol.* **2006**, *117*, 391–397. [CrossRef]
8. Worm, M.; Moneret-Vautrin, A.; Scherer, K.; Lang, R.; Fernandez-Rivas, M.; Cardona, V.; Kowalski, M.L.; Jutel, M.; Poziomkowska-Gesicka, I.; Papadopoulos, N.; et al. First European data from the network of severe allergic reactions (NORA). *Allergy* **2014**, *69*, 1397–1404. [CrossRef]
9. Umasunthar, T.; Leonardi-Bee, J.; Turner, P.J.; Hodes, M.; Gore, C.; Warner, J.O.; Boyle, R.J. Incidence of food anaphylaxis in people with food allergy: A systematic review and meta-analysis. *Clin. Exp. Allergy* **2015**, *45*, 1621–1636. [CrossRef]
10. Panesar, S.S.; Javad, S.; De Silva, D.; Nwaru, B.I.; Hickstein, L.; Muraro, A.; Roberts, G.; Worm, M.; Bilò, M.B.; Cardona, V.; et al. The epidemiology of anaphylaxis in Europe: A systematic review. *Allergy* **2013**, *68*, 1353–1361. [CrossRef]
11. Waserman, S.; Cruickshank, H.; Hildebrand, K.J.; Mack, D.; Bantock, L.; Bingemann, T.; Chu, D.K.; Cuello-Garcia, C.; Ebisawa, M.; Fahmy, D.; et al. Prevention and management of allergic reactions to food in child care centers and schools: Practice guidelines. *J. Allergy Clin. Immunol.* **2021**, *147*, 1561–1578. [CrossRef]
12. Alqurashi, W.; Stiell, I.; Chan, K.; Neto, G.; Alsadoon, A.; Wells, G. Epidemiology and clinical predictors of biphasic reactions in children with anaphylaxis. *Ann. Allergy Asthma Immunol.* **2015**, *115*, 217–223.e2. [CrossRef] [PubMed]
13. McIntyre, L. Administration of Epinephrine for Life-Threatening Allergic Reactions in School Settings. *Pediatrics* **2005**, *116*, 1134–1140. [CrossRef] [PubMed]
14. Pulcini, J.M.; Sease, K.K.; Marshall, G.D. Disparity between the presence and absence of food allergy action plans in one school district. *Allergy Asthma Proc.* **2010**, *31*, 141–146. [CrossRef]
15. Sicherer, S.H.; Furlong, T.J.; DeSimone, J.; Sampson, H.A. The US Peanut and Tree Nut Allergy Registry: Characteristics of reactions in schools and day care. *J. Pediatrics* **2001**, *138*, 560–565. [CrossRef]
16. Pouessel, G.; Lejeune, S.; Dupond, M.-P.; Renard, A.; Fallot, C.; Deschildre, A. Individual healthcare plan for allergic children at school: Lessons from a 2015-2016 school year survey. *Pediatric Allergy Immunol.* **2017**, *28*, 655–660. [CrossRef] [PubMed]
17. Alberta Laws. Available online: https://docs.assembly.ab.ca/LADDAR_files/docs/bills/bill/legislature_30/session_1/201905 21_bill-201.pdf (accessed on 22 February 2021).
18. Ontario Laws. Available online: https://www.ontario.ca/laws/statute/05s07?_ga=2.97067992.650686573.1614619401-1441028624 .1614013729 (accessed on 22 February 2021).
19. Food Allergy Canada. Available online: https://foodallergycanada.ca/professional-resources/educators/school-k-to-12 /national-school-policies/ (accessed on 18 September 2021).
20. Winnipeg School Division. Available online: https://sbwsdstor.blob.core.windows.net/media/Default/medialib/march-19-201 8.e6937d15056.pdf (accessed on 18 September 2021).
21. Ben-Shoshan, M.; Kagan, R.; Primeau, M.-N.; Alizadehfar, R.; Verreault, N.; Yu, J.W.; Nicolas, N.; Joseph, L.; Turnbull, E.; Dufresne, C.; et al. Availability of the epinephrine autoinjector at school in children with peanut allergy. *Ann. Allergy Asthma Immunol.* **2008**, *100*, 570–575. [CrossRef]

22. Shah, S.S.; Parker, C.L.; O'Brian Smith, E.; Davis, C.M. Disparity in the availability of injectable epinephrine in a large, diverse US school district. *J. Allergy Clin. Immunol. Pract.* **2014**, *2*, 288–293.e1. [CrossRef] [PubMed]
23. Hogue, S.L.; Muniz, R.; Herrem, C.; Silvia, S.; White, M.V. Barriers to the Administration of Epinephrine in Schools. *J. Sch. Health* **2018**, *88*, 396–404. [CrossRef]
24. Miles, L.M.; Ratnarajah, K.; Gabrielli, S.; Abrams, E.M.; Protudjer, J.L.; Bégin, P.; Chan, E.S.; Upton, J.; Waserman, S.; Watson, W.; et al. Community Use of Epinephrine for the Treatment of Anaphylaxis: A Review and Meta-Analysis. *J. Allergy Clin. Immunol. Pract.* **2021**, *9*, 2321–2333. [CrossRef]
25. Tsuang, A.; Demain, H.; Patrick, K.; Pistiner, M.; Wang, J. Epinephrine use and training in schools for food-induced anaphylaxis among non-nursing staff. *J. Allergy Clin. Immunol. Pract.* **2017**, *5*, 1418–1420.e3. [CrossRef]
26. Carlisle, S.K.; Vargas, P.A.; Noone, S.; Steele, P.; Sicherer, S.H.; Burks, A.W.; Jones, S.M. Food Allergy Education for School Nurses. *J. Sch. Nurs.* **2010**, *26*, 360–367. [CrossRef] [PubMed]
27. Hui, J.W.; Copeland, M.; Lanser, B.J. Food Allergy Management at School in the Era of Immunotherapy. *Curr. Allergy Asthma Rep.* **2020**, *20*, 32. [CrossRef]
28. Vale, S.; Smith, J.; Said, M.; Mullins, R.J.; Loh, R. ASCIA guidelines for prevention of anaphylaxis in schools, pre-schools and childcare: 2015 update. *J. Paediatr. Child Health* **2015**, *51*, 949–954. [CrossRef] [PubMed]
29. Page, M.J.; McKenzie, J.E.; Bossuyt, P.M.; Boutron, I.; Hoffmann, T.C.; Mulrow, C.D.; Shamseer, L.; Tetzlaff, J.M.; Akl, E.A.; Brennan, S.E.; et al. The PRISMA 2020 statement: An updated guideline for reporting systematic reviews. *BMJ* **2021**, *372*, n71. [CrossRef] [PubMed]
30. Ouzzani, M.; Hammady, H.; Fedorowicz, Z.; Elmagarmid, A. Rayyan—A web and mobile app for systematic reviews. *Syst. Rev.* **2016**, *5*, 210. [CrossRef]
31. Lawlis, T.; Bakonyi, S.; Williams, L.T. Food allergy in schools: The importance of government involvement. *Nutr. Diet.* **2017**, *74*, 82–87. [CrossRef]
32. Polloni, L.; Baldi, I.; Lazzarotto, F.; Bonaguro, R.; Toniolo, A.; Celegato, N.; Gregori, D.; Muraro, A. School personnel's self-efficacy in managing food allergy and anaphylaxis. *Pediatr. Allergy Immunol* **2016**, *27*, 356–360. [CrossRef]
33. Canon, N.; Gharfeh, M.; Guffey, D.; Anvari, S.; Davis, C.M. Role of Food Allergy Education: Measuring Teacher Knowledge, Attitudes, and Beliefs. *Allergy Rhinol.* **2019**, *10*, 2152656719856324. [CrossRef]
34. Shah, S.S.; Parker, C.L.; Davis, C.M. Improvement of teacher food allergy knowledge in socioeconomically diverse schools after educational intervention. *Clin. Pediatr.* **2013**, *52*, 812–820. [CrossRef]
35. Wahl, A.; Stephens, H.; Ruffo, M.; Jones, A.L. The Evaluation of a Food Allergy and Epinephrine Autoinjector Training Program for Personnel Who Care for Children in Schools and Community Settings. *J. Sch. Nurs.* **2015**, *31*, 91–98. [CrossRef]
36. Eldredge, C.; Patterson, L.; White, B.; Schellhase, K. Assessing the readiness of a school system to adopt food allergy management guidelines. *Wis. Med. J.* **2014**, *113*, 155–161.
37. Raptis, G.; Perez-Botella, M.; Totterdell, R.; Gerasimidis, K.; Michaelis, L.J. A survey of school's preparedness for managing anaphylaxis in pupils with food allergy. *Eur. J. Pediatr.* **2020**, *179*, 1537–1545. [CrossRef] [PubMed]
38. Ozturk Haney, M.; Ozbicakci, S.; Karadag, G. Turkish teachers' self-efficacy to manage food allergy and anaphylaxis: A psychometric testing study. *Allergol. Immunopathol.* **2019**, *47*, 558–563. [CrossRef] [PubMed]
39. Ercan, H.; Ozen, A.; Karatepe, H.; Berber, M.; Cengizlier, R. Primary school teachers' knowledge about and attitudes toward anaphylaxis. *Pediatric Allergy Immunol.* **2012**, *23*, 428–432. [CrossRef] [PubMed]
40. Rodríguez Ferran, L.; Gomez Tornero, N.; Cortes Alvarez, N.; Thorndike Piedra, F. Anaphylaxis at school. Are we prepared? Could we improve? *Allergol. Immunopathol.* **2020**, *48*, 384–389. [CrossRef]
41. Gonzalez-Mancebo, E.; Gandolfo-Cano, M.; Trujillo-Trujillo, M.; Mohedano-Vicente, E.; Calso, A.; Juarez, R.; Melendez, A.; Morales, P.; Pajuelo, F. Analysis of the effectiveness of training school personnel in the management of food allergy and anaphylaxis. *Allergol. Immunopathol.* **2019**, *47*, 60–63. [CrossRef]
42. Ravarotto, L.; Mascarello, G.; Pinto, A.; Schiavo, M.R.; Bagni, M.; Decastelli, L. Food allergies in school: Design and evaluation of a teacher-oriented training action. *Ital. J. Pediatr.* **2014**, *40*, 100. [CrossRef]
43. Polloni, L.; Lazzarotto, F.; Toniolo, A.; Ducolin, G.; Muraro, A. What do school personnel know, think and feel about food allergies? *Clin. Transl. Allergy* **2013**, *3*, 39. [CrossRef]
44. Polloni, L.; Baldi, I.; Lazzarotto, F.; Bonaguro, R.; Toniolo, A.; Gregori, D.; Muraro, A. Multidisciplinary education improves school personnel's self-efficacy in managing food allergy and anaphylaxis. *Pediatr. Allergy Immunol.* **2020**, *31*, 380–387. [CrossRef]
45. Gupta, R.S.; Kim, J.S.; Springston, E.E.; Smith, B.; Pongracic, J.A.; Wang, X.; Holl, J. Food allergy knowledge, attitudes, and beliefs in the United States. *Ann. Allergy Asthma Immunol.* **2009**, *103*, 43–50. [CrossRef]
46. Abrams, E.M.; Greenhawt, M. The role of peanut-free school policies in the protection of children with peanut allergy. *J. Public Health Pol.* **2020**, *41*, 2013. [CrossRef] [PubMed]
47. Bartnikas, L.M.; Sheehan, W.J.; Hoffman, E.B.; Permaul, P.; Dioun, A.F.; Friedlander, J.; Baxi, S.N.; Schneider, L.C.; Phipatanakul, W. Predicting food challenge outcomes for baked milk: Role of specific IgE and skin prick testing. *Ann. Allergy Asthma Immunol.* **2012**, *109*, 309–313.e1. [CrossRef] [PubMed]
48. Arksey, H.; O'Malley, L. Scoping studies: Towards a methodological framework. *Int. J. Soc. Res. Methodol.* **2005**, *8*, 19–32. [CrossRef]

MDPI
St. Alban-Anlage 66
4052 Basel
Switzerland
Tel. +41 61 683 77 34
Fax +41 61 302 89 18
www.mdpi.com

Nutrients Editorial Office
E-mail: nutrients@mdpi.com
www.mdpi.com/journal/nutrients